MISSION CONTROL

UNIVERSITY PRESS OF FLORIDA

Florida A&M University, Tallahassee
Florida Atlantic University, Boca Raton
Florida Gulf Coast University, Ft. Myers
Florida International University, Miami
Florida State University, Tallahassee
New College of Florida, Sarasota
University of Central Florida, Orlando
University of Florida, Gainesville
University of North Florida, Jacksonville
University of South Florida, Tampa
University of West Florida, Pensacola

MISSION CONTROL

INVENTING THE GROUNDWORK
OF SPACEFLIGHT

MICHAEL PETER JOHNSON

University Press of Florida
Gainesville · Tallahassee · Tampa · Boca Raton
Pensacola · Orlando · Miami · Jacksonville · Ft. Myers · Sarasota

This book may be available in an electronic edition.

20 19 18 17 16 15   6 5 4 3 2 1

Library of Congress Control Number: 2015938076
ISBN 978-0-8130-6150-4

The University Press of Florida is the scholarly publishing agency for the State
University System of Florida, comprising Florida A&M University, Florida Atlantic
University, Florida Gulf Coast University, Florida International University, Florida
State University, New College of Florida, University of Central Florida, University of
Florida, University of North Florida, University of South Florida, and University of
West Florida.

University Press of Florida
15 Northwest 15th Street
Gainesville, FL 32611-2079
http://www.upf.com

To my parents, David and Eileen, without whose tireless and selfless work, love, and support I could not have completed my goals

And to the thousands of anonymous men and women who worked in Mission Control to make humanity's greatest dreams reality

CONTENTS

ABBREVIATIONS

ASTP	Apollo-Soyuz Test Project
Caltech	California Institute of Technology
Capcom	Capsule Communicator, or Spacecraft Communicator
CERN	European Organization for Nuclear Research
COPERS	European Preparatory Commission for Space Research
DCC	Data Computation Complex
DCR	Dedicated Control Room
DCS	Display and Control System
DOD	Department of Defense
DSIF	Deep Space Instrumentation Facility
DSN	Deep Space Network
EAC	European Astronaut Centre
ECC	ESTRACK Control Centre
EECOM	Electrical, Environmental, and Consumables Manager
ELDO	European Launch Development Organisation
ESA	European Space Agency
ESDAC	European Space Data Acquisition Centre
ESOC	European Space Operations Centre
ESRIN	European Space Research Institute
ESRO	European Space Research Organisation
ESTEC	European Space Research and Technology Centre
ESTRACK	European Space Tracking Network
EVA	Extra-Vehicular Activity
FCR	Flight Control Room
FCT	Flight Control Team

FD	Flight Director
FOD	Flight Operations Director
GALCIT	Guggenheim Aeronautical Laboratory at Caltech
GEOS	Geostationary Earth Orbit Satellite
GSFC	Goddard Space Flight Center
ISS	International Space Station
JPL	Jet Propulsion Laboratory
JSC	Johnson Space Center
KSC	Kennedy Space Center
LEOP	Launch and Early Operations Phase, or Launch and Early Orbit
MCC	Mission Control Center
MCR	Main Control Room
MOCR	Mission Operation Control Room
MOD	Mission Operations Director
MSA	Mission Support Area
MSC	Manned Spacecraft Center
MSFN	Manned Space Flight Network
NACA	National Advisory Committee for Aeronautics
NASA	National Aeronautics and Space Administration
NASCOM	NASA Communications Network
OCC	Operations Control Centre
RKA	Russian Federal Space Agency
RTCC	Real-Time Computer Complex
SCOS	Spacecraft Control and Operations System
SFOF	Space Flight Operations Facility
SSR	Staff Support Room
STADAN	Space Tracking and Data Acquisition Network
STDN	Space Tracking and Data Network
STG	Space Task Group
STS	Space Transportation System, the Space Shuttle
TDRSS	Tracking and Data Relay Satellite System
UNIVAC	Universal Automatic Computer

MISSION CONTROL

INTRODUCTION

On 14 November 1969, Cape Canaveral, Florida, experienced rainstorms to the extent that the National Aeronautics and Space Administration (NASA) considered postponing the launch of *Apollo 12*. After much deliberation, NASA officials and meteorologists decided that the weather posed no threat to the vehicle. Blastoff occurred at 11:22 a.m. local time. Thirty-six seconds after launch, Commander Charles "Pete" Conrad noticed a flash outside the window. Static filled the communications. Sixteen seconds later, Conrad, along with lunar module pilot Alan Bean and command module pilot Richard Gordon, lost power to their guidance systems. Alarms clanged in their headsets and flashed on the control panels. Not even a minute into the mission, *Apollo 12* was doomed.

NASA did not realize until reviewing the tape that the spacecraft flying through the clouds acted like a conductor and had been struck by lightning twice. This overloaded the electrical systems, and the fuel cells shorted out. Fortunately for the astronauts, the backup batteries switched on and picked up the electrical load. If the batteries had failed, an automatic abort would have initiated. Unfortunately for them, they were flying blind, with no clear indication of how to fix the situation.

Meanwhile, mission control in Houston, Texas, continued to receive information from the *Saturn V* launch vehicle, but no data was conveyed from the command module. There was a possibility that this lack of data resulted from problems in the ground network, but that was ruled out when the astronauts confirmed that they had numerous alarms sounding and had lost all power, save emergency batteries. The situation looked dire, and flight director Gerry Griffin prepared to abort.

John Aaron, working the electrical, environmental and consumables manager (EECOM) console, recognized the errors he was reading from previous training. A year earlier, during a test of systems, his screens had registered nearly unintelligible values. Being naturally curious, he studied the test and contacted the simulation engineers to find out the source of the problem. With the help of other experts, Aaron eventually determined that the power had dropped from the module and could be fixed, with certainty, by an obscure switch called signal-condition equipment, or SCE.

The pattern in the data, therefore, caught Aaron's attention and sent him into action. He quickly told the capsule communicator controller to suggest that the astronauts change the SCE switch from "off" to "aux," or "auxiliary." This switch was so obscure that Conrad and Gordon did not understand the request. Fortunately, Bean at least knew of the existence of the SCE switch. After Bean completed the task, and the fuel cells reset, the systems came back on and *Apollo 12* was eventually allowed to complete its mission. The astronauts, taking a cue from their skipper, could not help but laugh the rest of the way into orbit. *Apollo 12* was saved, in part, because Aaron was the only controller who recognized the problem and the correct solution, and in part because, in another twist of fate, Bean was perhaps the only astronaut who knew the location and nature of the SCE switch.[1] Without this teamwork and communication, the second successful landing on the moon may have resulted instead in an embarrassing failure or worse.

* * *

Humans are social beings. They thrive on contact with other humans. The literature, both academic and fictional, on the effects of isolation on a human is copious and stunningly diverse. A human in space also requires a connection to humanity on earth. That voice on the other end of the communications link is mission control. More specifically, for NASA astronauts, the Mission Control Center of the Johnson Space Center has been that voice for nearly fifty years.

Mission control consists of more than just a voice. Mission planners and controllers develop flight plans for each mission. Controllers monitor the systems of the spacecraft as well as the bodies of astronauts themselves. In the event of a problem, the controllers consult with various engineers and other experts to find a solution and return the spacecraft to peak or near-peak working condition.

Any mission to space, human or robotic, represents years, if not decades, of work by thousands of individuals. Like the proverbial iceberg, mission control therefore is only the most visible part of the thousands of people on the ground attempting to ensure that the spaceflight is successful. Someone, whether scientist, engineer, or administrator, must advocate the initial plan for each mission. Mission planners organize the mission. Engineers design the spacecraft as well as its payload, or what it is carrying. Flight dynamics experts calculate the orbits and trajectories. Simulation engineers train and prepare the controllers. On manned missions, trainers prepare the astronauts. All of this must occur before the controllers even begin to communicate with the spacecraft.

MISSION CONTROL IN POPULAR CULTURE

If perception is reality, then examining the view of mission control in popular culture is a worthwhile exercise. If movies help shape popular opinion, then a look at how various movies have portrayed mission control will provide a window into the popular opinion of what mission control looks like and how it works. If the perception, therefore, tends to be substantially different from reality, then now is the time to enlighten the public about how mission control actually functions.

Most movies portraying a mission control tend to oversimplify the controllers themselves. Typically, one or at most two men talk to the astronauts in space, serve as the lead controller directing the room, and may even serve as an executive director for NASA, which often includes choosing and training the astronauts. The most prominent examples include Dan Truman (Billy Bob Thornton) from *Armageddon* and Bob Gerson (James Cromwell) from *Space Cowboys*. Even *Apollo 13*, arguably the most accurate depiction of mission control on film, tends to use Ken Mattingly (Gary Sinise) as a stand-in for all astronauts who worked in mission control and in the simulator during the famous rescue mission. Presumably, these all-in-one characters aid in story development by preventing an overabundance of non-astronaut characters, but this characterization does present a problematic and incorrect depiction in showing one individual wearing so many hats for the space agency. In reality, such characters are doing the job of at least a dozen different people.[2]

Similarly, all of NASA is typically reduced to two locations: Houston and Cape Canaveral. In *Space Cowboys*, Houston not only handles astronaut training and mission control, which is accurate, but also the monitoring of communication satellites, with astronauts involved. In *Armageddon*, the astronaut training in Houston includes not only the realistic underwater training at the Neutral Buoyancy Laboratory and T-38 flights out of Ellington Field, but also the hangar for the space shuttles and a vast desert area to test the Armadillos, the drilling vehicles to be used on the surface of the asteroid. Again, this can be rationalized as tight storytelling, but in some ways it cheapens the rest of NASA and further highlights the two most prominent centers, to the detriment and exclusion of the other NASA facilities.

Many movies have been able to film in select buildings on the campus in Houston, but none in the last forty years have been able to shoot in mission control itself. Even the most fictionalized movies, like *Armageddon*, have scenes in areas like simulators, office spaces, buildings, and even the large tank used for weightless training. Since each has had to construct its own control room, they vary in accuracy. *Armageddon*'s is clearly the most stylized, including television screens all over the room, a bank of monitors hanging from the ceiling, and an elevated console for the flight director. Others, like *Space Cowboys*, *Apollo 13*, and *From the Earth to the Moon*, constructed their control rooms down to small details.[3]

The precision of the depiction of mission control is directly linked to Hollywood's attempt for accuracy in the movies. Those, like

Armageddon, which are the most fantastical, typically have the least correct portrayals of mission control. As the movies move toward realism, so do their representations of mission control. Not surprisingly, those portraying real events, like *Apollo 13*, have nearly flawless mission controls.

Movies have depicted mission control in a variety of ways. Depending on which movie they have seen, the public may think of mission control, and all of NASA, as the domain of a handful of individuals or, correctly, as a truly massive undertaking. Most importantly, people need to realize that going to space is not easy, and it takes the efforts of thousands of people to accomplish that goal as safely as possible. Movies about any historical period or industry should strive to portray events in the most accurate way possible so as to avoid spreading misconceptions and even lies.

While there are positive interpretations of mission control, such as *Apollo 13* and *From the Earth to the Moon*, people should look to some of the documentaries made about NASA for the most accurate portrayal. In particular, a few History Channel presentations (*The Race to the Moon*, *Failure Is Not an Option*, and *Beyond the Moon: Failure Is Not an Option 2*) include insightful interviews with flight controllers and footage from the missions. Overall, these are important visual representations to complement the written accounts of space travel. Movies best serve history and the historical record when they strive for both an accurate set and credible characters. Most sectors of our history demand and receive this integrity of presentation; mission control demands no less.

SETTING THE RECORD STRAIGHT

For the majority of people, when they hear the term "mission control" they think of the Mission Control Center (MCC) at Johnson Space Center (JSC) in Houston, Texas. Houston's center is just one of dozens of mission control centers for spaceflight across the world. NASA has a handful of other control rooms in various centers around the United States. The Kennedy Space Center (KSC) in Cape Canaveral, Florida, houses the Launch Control Complex and previously included the Mercury Control Center. The Goddard Space Flight Center (GSFC) in Greenbelt, Maryland, maintains a control center for a number of satellites, including the Hubble Space Telescope. The Jet Propulsion Laboratory (JPL) in Pasadena, California, houses the Operations Control Center (OCC) for the Deep Space Network (DSN) and the majority of deep space missions. These are just the most prominent examples.

Outside NASA, most major space agencies have at least one, if not multiple, control centers. The main control center for the European Space Agency (ESA) is the European Space Operations Centre (ESOC) in Darmstadt, Germany. The European Space Research Institute (ESRIN) in Frascati, Italy, also includes a control room for ESA's Earth observation satellites. Human ESA missions have still another control room at the European Astronaut Centre (EAC) in Cologne, Germany. In addition, other national European space agencies have their own control centers.

Other space agencies have constructed their own control rooms. The main control center for the Russian Federal Space Agency (RKA) is in Korolyov, just outside Moscow. The Beijing Aerospace Command and Control Center serves as a control room for the China National Space Administration (CNSA). The Japanese Aerospace Exploration Agency (JAXA) maintains a control room at its Tsukuba Space Center in Tsukuba. The Indian Space Research Organization (ISRO) has a master control facility in Hassan. Numerous other control centers can be found across the globe.

A complete analysis of each mission control at this time is impractical. This work therefore focuses on the main control rooms for NASA at Johnson Space Center and the Jet Propulsion Laboratory and ESA at the European Space Operations Centre. These three centers provide an excellent examination of the two major types of spaceflight missions: human and robotic. They also allow for an examination of the differences between the space centers of different nations. This study can thus answer a few of the major questions surrounding the construction and use of mission control centers. How do mission control centers working with human spaceflight differ from those coordinating robotic spaceflight? How are domestic and international political issues reflected in the space programs in general and the control rooms in particular? What role did the Cold War have in the invention of mission control centers? And how did the control rooms adapt to the changing political landscape and the needs of the space agencies? These questions provide important focal points for this analysis.

A careful examination of the three mission control centers included in this project reveals that, while they developed independently, all three are more similar than they are different. This is most likely due to the fact that certain technologies and practices have proven to be most effective for controlling spaceflight missions. Where differences do arise, however, functionality overrides nationality as the most important determining factor. An outsider may believe that, above all, the nation or nations constructing and operating each mission control will have the most important imprint on how they are run. While this does have some influence, by far the most important aspect of how they function is the type of missions to be controlled. In the case of the control centers of this study, while JSC and JPL have some similarities, since they are both American, JPL actually operates more similarly to ESOC, the European robotic control center.

THE COLD WAR

Two other important concepts demand a brief introduction. Though it may be unnecessary to elucidate the obvious historical setting of space exploration, it is important to realize and remember that the backdrop for much of spaceflight history was the Cold War. It would be folly to re-create the vast historiography of Cold War history here, or to make a seemingly revolutionary claim about that time period. At least three key items, however, provide a reference point for the body of this work.

First, the Cold War was an ideological battle between the two world superpowers, the United States and the Soviet Union. While the two powers never engaged in a "hot" war, they did participate in a series of "proxy" wars, wherein they opposed each other through other means. The most notable of these proxy wars are the Korean War, the Vietnam War, and the Soviet War in Afghanistan. Space could be viewed as the venue for another proxy war.

After 1957, both the United States and the Soviet Union clearly viewed space as an essential aspect of their world power struggle. The propaganda value alone can be seen in the hero worship of astronauts like Yuri Gagarin, Valentina Tereshkova, Alan Shepard, and John Glenn. Following President Kennedy's declaration of the intention to reach the moon by the end of the 1960s, and the creation, whether real or perceived, of the space race, the United States set a clear victory point for Cold War technology. With space as the battleground, and the moon as the objective, both countries' space programs became proxy armies for the superpowers.

This leads to the final point about the Cold War: the centrality of technology. While the origins of the Cold War are many and complex, one of the most important aspects was the introduction of nuclear weapons to the countries' arsenals. Following World War II and the bombings of Hiroshima and Nagasaki, the United States stood as the undisputed military power with the most destructive potential. The Soviet Union moved as quickly as possible to catch up, and virtually did so with their first detonation of a nuclear weapon in 1949. Citizens around the world feared an apocalyptic nuclear war. Some kept a wary eye on the skies, some built fallout shelters, and some merely prayed that their worst fears would not come true. Cold War anxiety became a real epoch-defining issue.

Throughout the Cold War, technologies associated with nuclear destruction evolved into some of the most celebrated of the twentieth century. Jet engines made airline travel more accessible. Computers eventually became smaller and more ubiquitous. Rockets brought men into space and beyond. Thus, technology has an intimately complicated and an arguably essential relationship with the Cold War era.

CONTROL

The second concept necessitating introduction is "control." The term *control* has generated an increased amount of interest by scholars of history, sociology, and other similar fields of study and inquiry. Numerous case studies have been written about various types of control, from weights and measures to nuclear laboratories. Others have sought to define the term *control* historically—that is, how individuals and corporations have used it and its typical associations.

The contributors to *Cultures of Control* have argued that the idea of control, and indeed the word itself, is a relatively recent phenomenon. It is essentially always related with technological systems in mind. The creation of control rooms, probably originating in the twentieth century, changed the meaning from a more abstract concept to something with a corresponding physical space. Humans could create rooms with the latest and greatest technology, usually with electrical systems at their core, to manage the world around them.[4]

Another salient concept related to mission control is that, many times, the large-scale technological systems control not only the technology but also the people using the system. The people using the system typically choose to use it based on efficiency, ease of use, or other similar reasons. Once accepted, however, the user must conform to the technological system and can even be constrained by it. Thus the user has acceded control to the system.[5] Even when the controllers were forced to come up with solutions to problems onboard spacecraft, they were still constrained by the limits of the technological systems in use.[6]

Finally, the post–World War II world context cannot be ignored when discussing control. For one thing, following the war, the government became an essential element of advancements in large-scale command and control technologies. Perhaps even more importantly, as systems grew and began to encounter larger amounts of data, there came a point when an individual person could easily suffer information overload. Computers, at least since World War II, have become a necessary tool for technological systems. Without computing capabilities, humans could "drown" in the "flood" of information.[7]

These and other elements of control play a vital role in understanding mission control. The rooms help to quantify control by placing it within a specific space. Once in place, the control systems tend to manage not only the objects (i.e., spacecraft) but also the users (i.e., flight controllers). Finally, the mission control centers serve as viable case studies for postwar large-scale technological systems, since they are both created and maintained by governments and could not function without immense computing power.

Relating to control, technophobia has been the subject of countless novels, movies, and academic works. The origins probably stem from the Industrial Revolution and movements like that of the Luddites, when many people felt helpless in the face of drastic changes. While there are many examples from before World War II that could be used, for the purposes of staying within the same historical context this examination will focus only on works from the Cold War era.[8]

Between the Cold War, the nuclear age, and the space age, science fiction quickly became one of the most popular genres of the 1950s and 1960s. Nearly all science fiction from this time included some warning about the possibilities of unchecked technological development. The 1951 classic *The Day the Earth Stood Still* remains one of the most iconic examples of technophobia, in which a humanoid alien warns about the necessity of using technology for good rather than destructive purposes. A somewhat different example of Cold War anxiety surrounding technology comes from *The Invasion of the Body Snatchers* (1956). However the viewer interprets it, the filmmakers certainly were making a statement about the dehumanization of the Cold War by technology.

Whole franchises have been built around the fear of technology, particularly nuclear technology. A few examples include *Godzilla*, the *Terminator*, and the zombie subgenre. Even science fiction that shows hope for the future includes visions of technology's destructive potential. The *Star Trek* franchise long has served as a vehicle both to celebrate and warn about technology. While it heralds a future more accepting of racial and gender diversity, many villains, most famously the Borg, personify technophobia. Technology will ultimately become so strong and self-aware as to take away our humanity. Resistance, as they say, is futile.

*   *   *

With these and hundreds of other examples, there can be little doubt that technophobia has become nearly synonymous with science fiction after World War II and the introduction of nuclear technology. This is one reason why the idea of a mission control center could be so essential to the public. NASA, reaching to the moon and the planets, was fulfilling many of the milestones previously only imagined in science fiction. The next logical step would be the advancement of technology to a place where it could destroy humanity. Mission control on the ground, if nothing else, presented a sense of management of the situation for the public. NASA was monitoring not only its own spacecraft but also those of Cold War rivals. NASA and the other space agencies were regulating the technology, providing a potential sense of calm against the cacophony of technophobia.

The scope of this book includes the mission control centers at Johnson Space Center in Houston, Texas; the Jet Propulsion Laboratory in Pasadena, California; and The European Space Agency in Darmstadt, Germany. Each of these centers has contributed in their own way to the history of spaceflight. Each has had its own goals and its own trials and tribulations to overcome in order to fulfill its missions. Each has its own unique chronicle from conception to the present. Thus, it is best to understand the history and layout of each mission control, beginning with the Johnson Space Center, and noting the similarities and differences, before examining the broader concepts of mission control.

1

JOHNSON SPACE CENTER

The astronauts work diligently, despite their long time away from home. One calls Mission Control from the lunar module while the others retrieve geological samples of the moon. All seems to be normal, routine, even boring.

Suddenly a strange sight appears. The astronaut on the lunar module must be dreaming. There cannot be another living being out here. But then he sees another, and then a third. The beings ruthlessly kill the two astronauts who were taking the samples. The final astronaut panics and vainly attempts to leave the moon. It is too late. He and his spacecraft are destroyed.

In his distress, the astronaut called to Houston. Unfortunately for him, they were too far away to help. The three beings, the damage done, head to the nearby planet feeling like gods. They call the planet Houston, since that was the word used by the astronaut.

Thankfully, this is a work of fiction—*Superman II*, to be precise. Yet it provides an excellent and poignant example of the importance of Houston's Mission Control in NASA history, as well as in popular culture. For many people, Houston has become synonymous with spaceflight.

Since the creation of Project Mercury in the late 1950s, human spaceflight has been the most popular aspect of NASA's mission. As such, its Mission Control Center in Houston, Texas, has become the primary, or even the only, Mission Control in the public's eye. With human lives at risk during each mission, NASA had to be sure to create a control center that principally protected those lives.

When NASA created the center in Houston in the early 1960s, NASA was able to construct a Mission Control from the ground up exactly as it wanted. The Mission Control Center therefore may be perceived as the perfect image of Mission Control in the space agency's eyes. Thus, it is essential to understand the construction process of this center to fully analyze the ground segment of America's human spaceflight program.

CENTER HISTORY

The roots of the Mission Control operations group reach back to the oldest of all the NASA facilities, Langley Research Center (1917) in Tidewater, Virginia. NASA established the original Space Task Group (STG) at Langley shortly after the agency's creation on 1 October 1958. The National Advisory Committee for Aeronautics (NACA), NASA's predecessor (dating back to 1915), put one of its veteran aeronautical engineers and flight test experts, Robert Gilruth, in charge of STG. His mission was to develop a human spaceflight program for the United States to rival the Soviet Union's. When the STG began work on 5 November 1958, it consisted of only fifty people: thirty-five from the Hampton, Virginia, facility, and fifteen from the Lewis Laboratory in Cleveland, Ohio. The STG was tasked with continuing to develop and fly rockets for the newly formed Project Mercury.[1]

Although based in Virginia, much of STG's work took place at the Mercury Control Center in Cape Canaveral, Florida. In Florida, a few members of the STG—most notably Christopher C. Kraft, the first flight director; Tecwyn Roberts from Wales; and John Hodge from England—were mainly responsible for designing the original Mercury Control. Much of the concept came directly from Kraft, Gilruth's protégé in high-speed flight testing, who visualized the control room manned by experts on the spacecraft systems and the various aspects of the mission. This idea arose out of Kraft's experience with test flights, when a flight-test engineer on the ground monitored the flight and provided suggestions to the pilot. Before building the room, he sought the advice of various other experts, including test pilots and the new Mercury astronauts, especially Donald K. "Deke" Slayton. The idea of a control room came about largely due to the need to protect the astronauts. In the case of an emergency abort, for example, controllers on the ground would need to monitor various procedures quickly and relay them to the astronauts.[2] In this way, the engineers and experts on the ground could maintain a semblance of control over their spacecraft.

Crisis response also dictated the location of the original Mission Control. NASA chose Cape Canaveral as the site of its first Mission Control largely so that a controller on-site could monitor the rocket as it stood on the launchpad and during the critical first few moments of the launch. Because they could not always rely on the early radar and telemetry systems to provide accurate information, controllers watched the rocket through periscopes in case of an emergency and the need for abort arose. Early failed tests proved this to be shrewd foresight.

Technicians and STG controllers also manned remote stations around the world with equipment to communicate with the astronauts in space. These remote sites completed the majority of the actual monitoring of spacecraft. Certain kinds of decisions, however, had to be made by a central group of experts. Due to the limitations of the technology of the time, the remote sites could not send information and have it expeditiously processed at Mercury Control. Thus, Mercury Control was more of a hub, where controllers, the experts in their fields, made the most crucial decisions.

Of great importance, Mercury Control served as a significant training ground, a classroom of sorts, for many of the controllers who went on to work in Mission Control for future programs like Gemini and Apollo. Aside from Kraft, future flight directors Gene Kranz, John Hodge, and Glynn Lunney, among many others, initially worked in Mercury Control.[3]

The first mission controlled from Cape Canaveral was Mercury-Redstone 2, which included the first living being sent into space by NASA: Ham, a chimpanzee.[4] While some controllers had prior experience with test flights and rocketry, the space program was so new and different that it required a new set of procedures and expertise. Mercury Control essentially started from scratch, and the controllers methodically developed their jobs along the way.[5] The technology available to them was relatively crude, including a mechanical plotting board toward the front of the room based on estimations of the spacecraft's location. Only two controllers had access to visuals through a television set, the capsule communicator (capcom) and the flight director. Others relied solely on instrumentation readings.[6] These limitations in the technology of Mercury Control may have influenced the development of the capcom and flight director positions as special within the control room. Even after visual technology expanded to be available for all controllers, only capcom and the flight director had substantial communications with the astronauts in space. Regardless, the room and its technology proved adequate for the six Mercury manned missions.

By the end of Project Mercury, NASA realized that they needed to replace Mercury Control. The constant shuttling of personnel between Langley and Cape Canaveral wore on the members of the STG. Also, the facilities in Langley and the Cape were unable to cope with the needs of NASA's human spaceflight program. As it moved on to the Gemini and Apollo programs, NASA confronted increasingly difficult procedures in space, like rendezvous and extravehicular activities. Mercury Control was not equipped with the technology to carry out such maneuvers.

The human spaceflight program continued to grow as well, and neither Langley nor Cape Canaveral could manage the number of employees required for such exploits. Eugene Kranz also argued that while, at first, NASA thought that the control center should be near the hardware, they evaluated and quickly realized that it was more important to be near "feeder universities," particularly as new controllers with knowledge of computers were brought in for the Gemini program.[7] Veteran controllers had little knowledge of computers, so as the machines became increasingly important for Mission Control, they relied more heavily on college graduates who did have training in computers. NASA also decided that the human spaceflight program required its own administrative center, as well as a new control center.

In late 1960, NASA initiated its search for a location for a new human spaceflight facility. Parameters set by a committee helped to narrow down the list of possible locations. For instance, the site needed access to water transportation for large barges carrying rockets and rocket components, a moderate climate to avoid lengthy cessations of work, a nearby airport, an infrastructure of technical facilities and potential employees within a reasonable distance, an established infrastructure of higher education in close proximity, abundant electrical and water supplies, and at least one thousand acres of land at a reasonable price.[8] Even with such specific criteria, the site selection committee received dozens of applications. They soon narrowed down the search to twenty-three sites, including Jacksonville and Tampa, Florida; Baton Rouge, Shreveport, and Bogalusa, Louisiana; San Diego, San Francisco, Berkeley, Richmond, Palo Alto, and Moffett Field, California; four sites near St. Louis, Missouri; and Victoria, Corpus Christi, Liberty, Beaumont, Harlingen, and three separate sites in Houston, Texas. The selection quickly escalated into a political endeavor, with the congressmen and -women who represented the sites campaigning for their district. While the committee originally favored the site in Tampa, the Air Force decided not to close down the Strategic Air Command operations at MacDill Air Force Base for NASA's use. The committee, dismayed by this development, quickly decided on one of the Houston sites as the new primary choice.[9]

After an extensive search process, on 19 September 1961, NASA announced that the human spaceflight program would build a new Manned Spacecraft Center (MSC) for $60 million in Houston, Texas. The land, purchased from Rice University, included one thousand acres near Clear Lake, which feeds into Galveston Bay and the Gulf of Mexico.[10] Thus, large barges could easily navigate to and from the location. NASA also obtained the rights to work out of Ellington Field, an old air base from the world wars, only seven miles northwest of the MSC location. Nearby technical universities included Louisiana State University, the University of Texas, and Texas A&M University, among many others. Finally, the lack of harsh winter weather meant that operations would not be hindered. The location thus met each of the parameters established by the site committee.

Rice University officials were skeptical, understandably, about the prospects of human spaceflight. They included a clause in the purchase negotiations stating that if the human space program failed, the ownership of the land and infrastructure would revert to the university. Hence, most of the early buildings look utilitarian and are centered on a series of duck ponds, giving the inner area a campus-like feel.[11]

There is some debate as to the extent of political influence on the choice of Houston. According to James Webb, the NASA director, Lyndon B. Johnson, the Senate majority leader from 1955 to 1961 and then vice president, had little to do with Houston winning the bid. Webb vehemently maintains that NASA based the decision solely on the merits of the location. Individuals counterfactually reading into events had overstated the political angle.[12]

Lead flight director Chris Kraft, among many others, disagrees. He argues that political considerations played as much a role as any other factor. The chairman of the House Appropriations Committee, Albert Thomas, was a congressman from Texas's 8th district, which includes Houston. Vice President Johnson also hailed from Texas. According to Kraft, when Robert Gilruth met Webb, he argued that they should keep the STG in Virginia. Webb, not so subtly, made reference to Harry Byrd, a thirty-year veteran senator from Virginia, and his lack of enthusiasm for the space program.[13] Almost anyone with knowledge of national politics, congressional campaigning, and the disbursement of federal funds would find Kraft's account more credible. Politically motivated decision or no, human spaceflight operations moved to Houston.

Many personnel in STG did not receive the announcement about the move to Houston enthusiastically. For one thing, a large number of them had worked at Langley before the creation of NASA, so Virginia had been their home for years. With average temperatures in the nineties for three months of the year, excessive humidity making the heat even worse, and the penchant for hurricanes, given its location on the Gulf of Mexico, Houston had what few would call a "moderate climate." Certainly it could be considered more extreme than Tidewater, Virginia, which itself was hardly moderate in the summer. Complaining would change neither the move nor the overly enthusiastic welcome from the native Texans. Relatively inexpensive housing, as well as free or deeply discounted merchandise from some local vendors, offset at least some of the hardship.[14]

As part of their transition, the STG sent an advance team to Houston to set up temporary offices in the Gulfgate Shopping Center near downtown Houston at the intersection of Interstates 45 (the Gulf Freeway) and the 610 loop that encircles downtown Houston. This team reported back to Langley on the progress at the new location, sought out homes for the members, and generally prepared the STG members for their move to Houston. During the transition, NASA leased a series of buildings in Houston for temporary work spaces.[15] Human spaceflight operations slowly but surely transitioned to their new home in Houston.

The initial decision to move the STG to Houston did not necessarily include Mission Control. When deciding on the location of the Mission Control Center (MCC), NASA considered a number of factors. These included the site of the Gemini and Apollo project offices, the location of the flight operations division, the residences of the astronauts, the location of computer facilities, the availability of communications, and the knowledge of operations preparation. NASA soon narrowed down possible locations to Houston and Cape Canaveral. After deliberation, the location of the project offices and the astronauts made Houston the obvious choice.[16] Thus, on 20 July 1962, Webb officially announced that the Manned Spacecraft Center would house the MCC for future flights, beginning with *Gemini*'s.[17] Following the flight of *Gemini 3* in March 1965, Mission Control permanently moved to its new location in Houston.[18]

In its first three years, NASA spent about $240 million to construct the Manned Spacecraft Center (MSC). Almost half of that cost went to the construction of the Mission Control Center, the highlight of the new space center.[19] With the Mission Control Center came new communications technologies, which allowed the MCC to obtain all the spacecraft data from the network remote sites, rather than information remaining at those sites. The remote sites, therefore, quickly became obsolete, were disbanded, and MCC emerged as the centralized control area for human spaceflight.[20]

Shortly after former president and spaceflight advocate Lyndon B. Johnson died in 1973, Texas Senator Lloyd Bentsen sponsored a bill to rename the MSC after Johnson. Johnson himself had set a precedent by

Aerial view of JSC, 1989. Courtesy of NASA, http://grin.hq.nasa.gov/ABSTRACTS/
GPN-2000-001112.html.

giving the John F. Kennedy Space Center in Florida its name shortly after the president's assassination in 1963. On 17 February 1973, the MSC officially became the Lyndon Baines Johnson Space Center, or JSC. A formal dedication took place on 27 August 1973.[21] To avoid confusion, this work will refer to the space center in Houston as JSC, regardless of date.

Even after each of the spaceflight centers had been established in the locations for some time, domestic politics continued to play an important role in the centers' histories. For instance, in one of the great moments of political wrangling in spaceflight history, each of the major human spaceflight program centers in the United States fought over the right to be called the lead center for the space shuttle program during the late 1970s. Each had a compelling case. The Marshall Space Flight Center in Huntsville, Alabama, built and tested much of the hardware. The Kennedy Space Center had the primary launch and landing facilities and did some astronaut training. The Johnson Space Center included primary astronaut training and Mission Control. After much debate, JSC was named the lead center for the shuttle, thereby ensuring its primacy in NASA in the near future.

The NASA administration made this decision despite strong suggestions that with the advent of the shuttle MCC was no longer needed. The new spacecraft included enough computing power to fly on its own without the guidance of those on the ground. The flight controllers and operations personnel argued vehemently against this notion, asserting that Mission Control remained a vital ingredient for the success of all spaceflights. At the very least, they contended, it should remain in place for the first few flights as backup. While the flight controllers won the argument, the nature of the work carried out at JSC changed as it redefined its role for the shuttle program.[22]

One major change at JSC came with the creation of the Weightless Environment Training Facility (WETF). During the Apollo missions, the *Saturn V* had launched with such force that it created accelerations up to 15 G's on the human body, an acceleration up to fifteen times that of gravity. In order to prepare for the launch, NASA had had to build a centrifuge in Houston to subject the astronauts to such extreme forces. The shuttle, however, created only about 3 G's of force, or roughly equivalent to a roller coaster. Consequently, NASA no longer needed the centrifuge. In its place, JSC created the WETF, essentially a large pool used to simulate the near-zero gravity of space. Engineers placed mockups of space hardware in the WETF so the astronauts could practice EVAs, or spacewalks.[23] Adaptation was crucial to the continuing viability of JSC.

By the mid-1990s, as the reality of a large space station became more apparent, JSC realized the WETF was too small to hold all the mockup components. In response, they built the Neutral Buoyancy Laboratory (NBL), later named the Sonny Carter Training Facility, off-site at the nearby Ellington Air Force Base. The tank is 202 feet (62 meters) long, 102 feet (31 meters) wide, 40 feet (12.34 meters) deep, and holds 6.2 million gallons (23.5 million liters) of water. When it was completed in 1997, it was large enough for two simultaneous simulation activities, though the space station's growth has been sufficient to outgrow the tank.[24]

As it adapted to new challenges, JSC incorporated a range of activities and duties in its many support facilities. Building 1 on the campus was the administration building, including offices for the JSC director and many of the other branch directors and managers. Building 2 consisted of an auditorium, the public affairs office, and the original visitor's center. Building 4 housed the astronaut offices. Building 5, the Jake Garn Mission Simulator and Training Facility, included dynamic simulators that can move to replicate launch and landing. Building 9, the Space Environment Simulation Laboratory, held mockups of the International Space Station, a full-scale mockup of the shuttle, crew compartment mockups of the shuttle, and a training area for the robotic arms. Building 14 housed the anechoic chamber, which simulated the quiet of space. Building 17 accommodated the kitchen, where specialists prepared food for spaceflight. Building 29 housed the centrifuge before NASA repurposed it to create the WETF. Building 30 was Mission Control Center. NASA stored many of the moon rocks in laboratories in building 31. Building 32 housed one of the largest vacuum chambers in the world. Spacecraft and components were taken to building 49, the Vibration and Acoustic Test Facility, to test their ability to withstand the vibrations and sounds of launch. All of these buildings played a vital role in flight control or astronaut training, the two main elements of one of NASA's leading centers.

MISSION CONTROL CENTER

NASA realized that it needed a new Mission Control Center for the Gemini and Apollo missions during the Mercury Program. The Mercury Control Center did not have enough capability to conduct rendezvous maneuvers, let alone anything more complex.[25] Thus, NASA built the Mission Control Center (MCC) at the new Manned Spacecraft Center in Houston, Texas.

NASA originally constructed the Mission Control Center (building 30) in two major stages. With an almost $800,000 contract, the Peter Kiewit Sons Company built the foundation and structure, finishing on 29 May 1963. The Ets Hokin and Galvin Company completed the rest of the major work on the building. In all, NASA paid roughly $8 million in contracts to complete the MCC.[26]

The three-story, 90,000-square-foot building had three distinct wings. The main entrance in the lobby wing connected the Mission Operations Wing (MOW) to the west and the Operations Support Wing (OSW) to the east. As the name suggests, the MOW included the Mission Control Rooms, the adjoining Staff Support Rooms (SSR), and the computer complex. NASA renamed the SSRs Multipurpose Support Rooms (MPSR) for the shuttle missions. During Gemini, Apollo, Skylab, and ASTP, when astronauts returned to earth by splashing down in the ocean, this wing also included a recovery control room. This provided an area for the Department of Defense (DOD) and NASA personnel to coordinate recovery of the astronauts and their spacecraft. This wing was windowless in an attempt to make it weatherproof and to limit interruptions from outside radio waves. The windowless environment, combined with a seeming myriad of identical hallways, can be confusing even to the most seasoned flight controller, or downright labyrinthine for a visitor.[27] The OSW included offices for the controllers and support staff.[28]

NASA held a groundbreaking ceremony for building 30 in April 1962. The building included the control rooms, computer complex,

Houston MCC, 2011.

and support rooms.[29] The MCC, as originally conceived, consisted of three major technology systems. The Communications Interface System (CIS) focused on communications both within the MCC and with certain outside areas, such as the NASA communications network (NASCOM) and simulators. The data computation complex (DCC) included the mainframe computer system. Finally, the display and control system (DCS) handled human interface within the MCC. The majority of data, therefore, entered the MCC and was distributed through the CIS. The DCC then processed the data, and the DCS displayed it.[30]

Like most control systems, the DCC and its mainframe computers were the "brains" of Mission Control. Virtually all data traveling to and from the MCC was processed by those mainframe computers, and then (working like the nervous system) sent to the appropriate location for use. The mainframe computers were periodically updated in order to keep Mission Control current with the best technology.

The MCC is now 102,000 square feet, providing a significant amount of space for the Mission Control rooms as well as support rooms and offices. It originally included Mission Operation Control Rooms

(MOCR) on the second and third floors. The third-floor MOCR ultimately controlled forty-two flights, including *Gemini 4–12, Apollo 4, 6, and 8–17,* and twenty-one space shuttle flights, most of which were classified Department of Defense missions. The second-floor MOCR controlled 114 missions, including nine Gemini missions, seventeen Apollo missions, Skylab, the Apollo-Soyuz Test Project, and eighty-three space shuttle flights.[31] Since NASA continually updated the control rooms, they traded off primary use during upgrades.

The MCC also housed the Network Interface Processor (NIP), located on the first floor. The NIP distributed incoming digital information to the needed areas of the MCC. Also on the first floor, the Data Computation Complex (DCC) compared incoming telemetry information with predictions, checking for anomalies. NASA built a Payload Operations Control Center (POCC) for space shuttle payload operations. The POCC included a control room, mission planning room, and six support rooms, totaling four thousand square feet.[32] All of these support areas hosted critical elements of control for manned spaceflight.

Nearby, building 48 housed generators for use in the event of a power failure, a potentially disastrous problem for Mission Control. It also provided an air-conditioning system for building 30.[33] In the event of a catastrophic failure, the White Sands Test Facility in New Mexico included an emergency backup control room.[34] JSC prepared for any emergency, with backups and redundancies similar to the spacecraft they oversaw.

At the time of its construction, the MCC boasted the largest amount of television switching equipment worldwide. As part of that system, it included 136 television cameras and 384 television receivers. The building utilized fifty-two million feet of wiring.[35] This allowed controllers and other employees not working inside the MOCRs to monitor the missions as they flew.

The control room and its equipment required a massive amount of computing capability. A Univac 490 computer originally ran communications as well as telemetry and trajectory information. The Univac 490 had 128 kilobytes of memory and one megabyte of head drum storage. For comparison, the iPhone 3GS has eight gigabytes of memory, or about sixty-five thousand times as much memory.[36] The mainframe included five IBM 7094s. These were located in the Real Time Computer Complex (RTCC).[37] The RTCC processed all data and telemetry information for missions controlled in the MCC.[38] NASA originally contacted ninety-four companies on 21 March 1962 to gauge their interest in submitting a proposal to build the RTCC. Of those, twenty responded with interest.[39] Eleven companies eventually bid to build the RTCC, including the Burroughs Corporation, the Control Data Corporation, the General Electric Company (GE), International Business Machines Corporation (IBM), International Telephone & Telegraph Company (ITT), Lockheed Aircraft Corporation, Philco Corporation, Radio Corporation of America, the Raytheon Company, System Development Corporation, and the Wilcox Electric Company. After much deliberation, the search committee stated that three of the proposals were clearly the best—those of IBM, ITT, and GE. IBM already held the contract for the Mercury Control Center. It also promised better organization as well as a more favorable cost estimate. Chris Kraft aided IBM's cause by campaigning for them as well. IBM, not surprisingly, won the contract, which included design of the RTCC and its implementation, costing more than $36 million. NASA and IBM signed the contract on 16 October 1962.[40]

NASA then upgraded the computers before the Apollo program. They replaced the Univac 490 with a Univac 494 and the IBM 7094s with IBM 360/75s in 1967.[41] For the shuttle, JSC installed IBM 370/168s in 1976 to take the place of the IBM 360/75s. In 1983 they added an IBM 3081 for other processes. A 1986 upgrade brought in four IBM 3083/JXs to replace the 370/168s. Three years later, another upgrade included IBM 3083-KX machines.[42] The mainframe computers soon proved not to be cost-effective, since almost eighty personnel were needed to staff the outdated machines during shuttle missions.[43]

It should be noted that not all of the flight controllers welcomed new technologies, especially the introduction of computers. This was not necessarily because they were unfamiliar with the new technology, although that may have influenced their feelings. Mainly they worried that the computers would be yet another piece of hardware that could fail, resulting in loss of data or, worse, loss of communication with the astronauts.[44] Like many Americans in the Cold War, controllers viewed technology as necessary but potentially dangerous. For flight controllers, loss of data or communication was considered a worst-case scenario. Even among people most dependent on technology, a certain amount of technophobia may arise.

In the 1990s, NASA realized that the old MOCRs could no longer adequately handle space shuttle and planned space station operations. They built three new Flight Control Rooms (FCRs) in the southern wing of the MCC. One conducted shuttle missions, another monitored the space station, and the third served as a training facility. With the new FCRs came new hardware.

JSC installed Loral Instrumentation 550s with IBM RISC-6000 computers for telemetry.[45] Rather than using mainframe computers, they integrated two hundred Digital Equipment Corp. workstation computers. One hundred thirty thousand feet of fiber-optic cable created the world's largest fiber-optic local area network. The amount of memory space grew exponentially as well, including 190 gigabytes of data storage.[46] This is more than one and a half times more memory than that of the original UNIVAC 490.

A series of Staff Support Rooms (SSR) surrounded the operations room. Specialists occupied these rooms for the various aspects of missions and were backups for the controllers in Mission Control.[47] MCC also included a Spacecraft Analysis Room (SPAN), generally acknowledged as one of the most important support rooms. SPAN housed senior engineers and controllers representing some of the major contractors for various aspects of the given mission. In essence, if a problem occurred, the flight controllers contacted SPAN. Those in the room then pared the information down to a specific question requiring an answer. The senior contractor members then consulted with their constituents for a solution. In some ways, SPAN served as an elaborate dispatcher.[48] For Skylab, NASA renamed the room the Flying Operations Management Room (FOMR).[49] After Skylab, it was renamed SPAN.[50]

Along with the contractors, SPAN communicated with another room in building 45 of JSC, the Mission Evaluation Room or MER. While SPAN identified the anomaly, MER did much of the work to solve the problem. As a result, many of the most knowledgeable and skilled engineers congregated in this room. It was so important, in fact, that many of the engineers considered the MER a step above Mission Control. Instead of relying on advanced technology, MER depended on its collective brainpower.[51] Both SPAN and MER proved invaluable during missions, often providing necessary solutions for anomalies while the controllers focused on other issues.

NASA installed sleeping quarters for the flight controllers near Mission Control. During the early Gemini missions, the flight controllers generally stayed on campus and slept in the designated rooms between their shifts so that they would not have to spend time commuting. After the Gemini program, however, the sleeping quarters were rarely used, except in extreme cases. The rooms simply were neither quiet nor comfortable, and the controllers realized the need for rest, especially during the stressful lunar missions.[52]

MISSION OPERATIONS CONTROL ROOMS TO FLIGHT CONTROL ROOMS

Cape Canaveral hosted the first control center for NASA's manned spaceflight program. The Mercury Control Center acted as a central location for operations, though much of the actual control work occurred in remote sites across the world. NASA realized that the more advanced missions of Project Gemini and the future Apollo Program necessitated a more permanent and centralized control center. The result was the Mission Control Center (MCC) in Houston, Texas.

Of the lessons learned from Mercury Control before building the Mission Operation Control Rooms (MOCR), one of the most important was the need for flexibility. NASA built Mercury Control solely to control the one-man missions. They configured each of the consoles to suit the individual controllers. They realized that this style of control room would not work well for the more permanent rooms needed by an unknown number of controllers for at least the Gemini and Apollo programs.[53]

Houston's MOCR had an unexpected first experience while controlling a mission in January 1965. During liftoff of *Gemini 2*, a power failure occurred in Mercury Control at Cape Canaveral. When the backup controllers stationed in Houston could not hear Mercury Control, they began to track the Titan rocket. Because power could not be restored at the Cape until reentry, the Houston control center effectively controlled the entire mission.[54] The MOCRs did not officially take over primary control until *Gemini 4.*

The MCC in Houston was fully operational in time for the June 1965 flight of *Gemini 4.* This flight secured its place in history when Ed White became the first American to perform an extravehicular activity (EVA), commonly referred to as a space walk. Project Gemini served largely as a training ground for the missions to the moon. In all, ten missions successfully flew between 1965 and 1966. NASA felt ready for the next step.

The Apollo program provided some of the most celebrated moments in NASA's history. It began, however, with one of its most tragic events. On the night of 27 January 1967, during what was supposed to be a routine test, a fire broke out in the command module of *AS-204,* later renamed *Apollo 1,* the first scheduled manned mission of the Apollo program. Unable to escape, three astronauts, Virgil "Gus" Grissom, Ed White, and Roger Chaffee, died of asphyxiation. After months of investigation, a review board determined that the spacecraft contained too many flammable materials in a pure oxygen environment. A spark from a frayed wire had caused some of those objects to catch fire. NASA set out to fix the many problems found in the command module. Numerous NASA employees mark the fire as a turning point for NASA—a loss of innocence as well as an awakening of NASA community from any sense of invulnerability. Indeed, many see NASA's history as prefire and postfire.

After the *Apollo 1* fire, Flight Director Gene Kranz defined Mission Control in two words: tough and competent. Flight controllers were tough by accepting their responsibilities and accountability for their actions. They were competent by understanding their role and never taking it for granted. Those two words remain the calling card of Mission Control.[55] Two other words added at other times were: discipline and morale.[56]

Since its inception, Houston has taken control of a spacecraft as soon as it "cleared the tower." While there are a few explanations for this, one seems to be the most logical. Launch Control at Kennedy Space Center, at least for the early Gemini and Apollo launches, had a periscope to view the launch. In the event of a launch failure, as had happened somewhat frequently with early unmanned launches, the launch director had an abort button. One possible failure was contact with the tower. As technology improved, this direct visual became less important, but the idea of handover after clearing the tower remained. Thus, there remains a Launch Control on-site in Florida, but Mission Control in Houston takes over primary control mere seconds after launch.[57]

Each MOCR is approximately 7,800 square feet.[58] Compared to the previous Mercury Control, the MOCRs, as Kraft described them, were spacious with faster computers and impressive support rooms nearby to keep the controllers abreast of the data transmitting from the spacecraft.[59] Philco Corporation constructed the equipment for the room, including cables and pneumatic tubes, under NASA contract NAS 9-1261. Philco had previously served as the major contractor for the Mercury Control Center; so, like IBM, it had an inside advocate in Chris Kraft. The original contract from 1963 cost more than $35 million.[60]

Controllers referred to the room's lighting as "a kind of perpetual dusk." Keeping the lights low aided the viewing of the console screens.[61] Three ten-by-twenty-foot screens covered the front of the room. These screens typically showed important data on the left, a world map for tracking purposes in the center, and any live feed from the mission on the right. The system used rear-projection equipment located in a dark room behind the screens, known as the "bat cave." In 1989, NASA carefully replaced the original glass screens, which weighed twelve hundred pounds. These screens highlighted one of the most important changes from the Mercury Control Center, namely, computerization. The front of the room also displayed the mission clock, the most precise measure of the mission duration.[62] Previously, controllers had tracked the missions using mechanical plot boards. The mainframe computers associated with the new MOCRs allowed the color map to display the tracking information electronically.[63] John Hodge, one of the first flight directors and designers of the MOCR, insisted that the screens at the front were merely for publicity. Visitors to the control room enjoyed looking at them, and they created a certain flair for the room, but the controllers themselves never used them. Instead they relied on their own consoles and their support rooms.[64]

Mission plaques hang on the side walls of the MOCRs. The room controlling a particular mission had the privilege of displaying the plaque for that mission. At the end of each mission, the controllers had a ceremony during which a flight controller hung the plaque. The flight directors also selected an individual honoree who had distinguished him or herself in some way during that mission.[65]

The viewing area behind the controllers was reserved for astronaut families, dignitaries, and other invited guests during missions. It seated seventy-four people.[66] This was also the closest to Mission Control that the vast majority of people come. It has been said that every president from Lyndon Johnson to Bill Clinton visited either the control room itself or the viewing area (George W. Bush visited as governor of Texas).

Some have described the camaraderie among the flight controllers as similar to that of a combat unit, due to a friendship borne out of dependency on one another to do the job, that is, to complete the mission successfully.[67] The military background of many of the early controllers made this analogy particularly apropos. Chemistry among the flight controllers was especially important. The more they worked together the more they anticipated each other and communicated using an economy of words.[68] The controllers had complete trust, respect, and confidence in each other.[69]

Many of the controllers were contractors from different companies related to their position. Others were NASA employees. In fact, during the Gemini and Apollo programs, nearly all of the MOCR controllers were NASA employees, whereas contractors staffed the support rooms. The flight director (FD, or Flight) was always a NASA employee.[70] This difference between contractors and employees could sometimes hurt the camaraderie of the room, though the need to complete the mission often overcame any differences.

The flight director truly was the center of Mission Control. The FD console stood in the center, and the other controllers surrounded him like concentric circles around a focal point. Outside the room, Staff Support Rooms (SSR) surrounded Mission Control, further adding

Flight director console, 2011.

to the image of concentric circles of control with the flight director at the center. The flight director had a rather simple yet complex job description: to "take any action necessary for crew safety and mission success."[71]

The flight director was at all times responsible for every aspect of the mission. Any mistake could be fatal to the astronauts, and he or she would shoulder the blame. Before being thrust headlong into the combat zone that is Mission Control, however, each flight director completed hundreds of simulations that included more errors than one might fear could ever happen in a real mission. This experience was vital, not only in the case of an emergency, but also, if nothing else, to prove that each flight director was ready for the immense responsibility. Flight directors, like all controllers, fought the tendency to revert to a routine. Maintaining a sense of awareness and avoiding apathy were critical aspects of successful spaceflight.[72]

Flight controllers began by working simulations in the "back rooms." Each worked various positions in teams during yearlong rotations, similar to medical students' rotations. After some years of gaining experience, they made their way up to the front room, again honing their skills in simulations before going on a true mission. If they were good enough and stayed on the job long enough, they gained the opportunity to advance to the most desired position: flight director. In 2009, JSC named three new flight directors, Dina Contella, Scott Stover, and Ed Van Cise, bringing the historical total to eighty.[73] With only eighty flight directors in forty-five years, however, only a select few ever received the coveted call sign "Flight."[74] As Chris Kraft, the first flight director, succinctly stated: "Flight is God."[75]

The amount of power and responsibility given to the flight director stands out as unique among spaceflight. The other control rooms included positions with some semblance of an FD, but none had the overall final say as at JSC. The exceptional amount of accountability inherent in manned spaceflight probably required such a role, and the men and women who have served as FD have deserved the acclaim they have received.

Shortly after the move to Houston, NASA hired a new crop of flight controllers directly out of college. Many were needed to operate the new computers installed in Mission Control.[76] Largely due to this influx of young controllers, in 1965 the average age for a flight controller was twenty-nine.[77] As NASA added more recent college graduates during the Apollo program, the average age declined to twenty-six in 1969.[78] The youngest flight controller ever was Jackie Parker, who was only eighteen when she joined Mission Control in 1979 as support for the data processing systems (DPS) console.[79] For the majority of the time, Mission Control, like most high-stress workplaces, has had a relatively low average age. As controllers gained more experience, they generally moved on to more senior positions.[80]

Early on, Mission Control strove to gather the best group of individuals possible, regardless of race, religion, or any other descriptive. In fact, Gene Kranz described it as one of the first true equal-opportunity government employers.[81] While it was an all-male environment for its first five years, the first women joined Mission Control as flight controllers in 1971.[82] It should be noted that three women worked in the mission planning and analysis division in the 1960s, a group that worked closely with Mission Control. At least one of those women,

Anne Accola, worked in the MOCR during *Apollo 17*, though not as an official member of the flight control team.[83] Today, approximately 40 percent of flight controllers are women.[84] NASA hired the first African-American flight director, Kwatsi Alibaruhu, as part of a new class of nine in 2005, the second-largest class ever, which brought the total at that time to thirty.[85] The class of 2005 also included the first two Hispanic flight directors: Ginger Kerrick and Richard Jones.[86]

Before each mission, flight directors were given certain tasks for that particular mission. While they each worked set shifts, there were typically special teams for launch, landing, and any other critical aspects of missions, such as lunar landing. Each flight director then assembled the best flight-control team for his or her particular task. As a result, teams at times remained relatively constant under a certain flight director, but there was still the possibility of change before each mission.[87]

Simulations were vital for the flight directors to understand how their team worked together. Some controllers became so involved and focused on their own work that they did not realize it when anomalies were occurring elsewhere. The flight director needed to instill a sense of teamwork and understand that problems could only be solved if the team were working in unison.[88] Thus one of the most important aspects of the flight director's job was creating strong chemistry among team members, something especially important during the Gemini and Apollo days, with so many young controllers straight out of college.[89]

The early flight directors each had a designated color that then provided a name for their team. For instance, Gerry Griffin was gold, Gene Kranz was white, Cliff Charlesworth green, Glynn Lunney black, Milton Windler maroon, Charles Lewis bronze, Neil Hutchinson silver, Don Puddy crimson, and Phil Shaffer purple. Jay H. Greene became emerald flight, because green already belonged to Charlesworth.[90] The flight directors often provided some flair for their teams. Kranz's wife made a vest for him to wear for each mission. A particularly colorful vest worn at the end of each mission designated his approval of their work. Puddy had a tendency to wear polka-dot shirts, and Shaffer striped shirts.[91] During the shuttle era, NASA ran out of colors for the flight directors and so ended this tradition.

Mission Control has had a few mascots over the years. During

Skylab, Lewis's bronze team adopted "Splash Gordon," a fish onboard the space station, while Hutchinson's silver team adopted Arabella, Skylab's spider.[92] One of the most memorable was Captain Refsmmat, named after a term used to describe equations to determine angles with reference to certain stars. The controllers hung an image of their mascot in the MCC, where they could anonymously write down any issues they were having with the mission. This "ideal flight controller" allowed the controllers to let off some steam and provided a sense of unity among the teams. It was further seen as a device to boost morale, a harkening to the military background of many of the early control-lers.[93] Some elements of Mission Control tried to start rival mascots, such as Victor Vector and Quincy Quaternion, but neither gained the popularity of Captain Refsmmat.[94]

The location of flight-control positions has changed somewhat, de-pending on the program or even the particular mission. Overall, Mis-sion Control consisted of a series of rows of consoles, each on a slightly higher level than the one before it, moving from the front of the room to the back. The following summary provides a general understand-ing of the various console positions throughout the course of Mission Control's history.

The first row, or "trench," included various positions concerned with the mechanics of the spacecraft. From Gemini to ASTP, the first posi-tion of the trench was the booster systems engineer (BOOSTER), who was responsible for the rocket stages. The booster controller came from the Marshall Space Flight Center in Huntsville, Alabama, where they built the rockets. Shortly after launch and the separation of the booster stages, the booster left the room, further adding to the position's dis-connection from the rest of the MOCR. The flight dynamics officer (FDO or FIDO) remained relatively constant. FIDO oversaw the ve-locity and trajectory of the spacecraft during all aspects of the mis-sion. Due to the time-critical aspects of their position, FIDO was the only position in Mission Control, other than the flight director, who could abort the mission directly. Any other controller had to secure an abort through the FD.[95] During Gemini and Apollo, FIDO served as the leader of the trajectory team, which also included the retrofire officer (RETRO) and the guidance officer. Before the shuttle program,

MOCR view from trench, 2011.

JSC decided that only one trajectory controller was needed, so they ended the RETRO position.[96] JSC merged the guidance officer (Guido) with the rendezvous procedure officer to create the guidance procedure officer (GPO, though still called "guidance"). The GPO was concerned with the positioning of the spacecraft and any deviations from the projected location. The guidance, navigation, and controls systems engineer (GNC) moved around in the MOCR somewhat, starting in the first row but shifting to the second row during early shuttle missions, before moving back to the first row. GNC covered navigation and some propulsion aspects in maneuvers. The propulsion engineer (PROP), created out of a reorganization before the space shuttle program, controlled all but the main engines for the shuttle.

The second row received the moniker "systems." During the Apollo program, consoles on the left included the flight surgeon and capcom. Capcom was the only controller to communicate verbally with the astronauts in space on a regular basis. The name originally stood for capsule communicator, but after the shuttle it was officially known as the spacecraft communicator. Capcom was always manned by an astronaut, because capcom served as that vital link between those on the ground and those in space who were familiar with all of the jargon affiliated with space travel. They even served as a sympathetic interme- diary between the controllers and the astronauts in space.[97] The idea started with the remote communication sites during the Mercury pro- gram. The astronauts flew out to those remote sites and served as the communications liaison with whomever was in space. For the shuttle, the surgeon moved to the back row and the capcom moved to the third row. The data processing system engineer (DPS) handled the onboard computers since the first flights of the shuttle. Another shuttle posi- tion, the payloads officer (PAYLOADS) served as a liaison between the groups responsible for the payload, usually a contractor or scientific experimenter and Mission Control.[98]

Other positions on the second row during Apollo dealt mostly with communications. Those included the electrical, environmental, and communications systems engineer (EECOM), the telemetry, electri- cal, EVA mobility unit officer, (TELMU, originally TELCOM), which was concerned with the electrical and environmental systems of the lunar module (LM) and spacesuits, and CONTROL, which handled the communications systems of the LM. JSC eventually consolidated these positions or reformed them into other positions. The electri- cal generation and illumination engineer (EGIL, pronounced "eagle") oversaw the power system on the shuttle. JSC created it out of some aspects of EECOM. With the shuttle, EECOM stood instead for the environmental engineer and consumables manager, who assured vi- ability of the life-support systems and consumables. Apollo EECOM flight controller Charles L. Dumis likened the systems controllers to plumbers. People hardly paid attention to them until something broke, and then it was their job to fix it.[99]

The third row handled many of the command aspects of missions. The integrated communications officer (INCO) managed the com- munications links with the spacecraft in space. During Gemini and Apollo, the operations and procedures officer (O&P) tracked displays and the mission clock. At times, an assistant flight director position served to aid the FD with some of the administrative duties. A contro- versial position, it functioned as assistant to the flight director.[100] The

MOCR back rows, 2011.

flight director had a console in the center of the third row. The flight activities officer (FAO) managed the astronauts' schedule as they were flying. The network position was concerned with ground communications. For the shuttle missions, NASA consolidated it and O&P into the ground controller, or GC.

The final row handled much of the liaison work for Mission Control with various outsiders. The first console in the back row rotated depending on the mission and the current aspect of the mission. Another shuttle controller was the mechanical, maintenance, arm, and crew systems engineer (MMACS, pronounced "max"). The MMACS controlled the robot arm, auxiliary power, hydraulic systems, payload bay doors, and various other mechanical systems. This position was likened to an all-around mechanic for the space shuttle. The booster engineer managed the main engines of the shuttle as well as the solid rocket boosters during shuttle launches. Previously, the position had monitored the engines and propellant tanks for the rockets. The booster position was manned only during the launch phase. One more position, added for shuttle operations, was the payload deployment and retrieval system

(PDRS) specialist. PDRS oversaw operations of the remote manipulator system, the robot arm of the shuttle. The extravehicular activity (EVA) specialist worked on console during all space walks.

Flight directors sometimes moved on to the mission operations director (MOD) position. The MOD essentially served as a liaison between Mission Control and the outside world. Before 1983, JSC called this position the flight operations director (FOD). A medical doctor stationed in Mission Control, called the flight surgeon, was the only console position outside of capcom that had regular communications with the astronauts in space. Flight surgeons performed medical checks on astronauts, both prior to and after missions, as well as during the mission itself. The public affairs officer (PAO) was the "voice" of Mission Control. The PAO interacted with the news media, handled publicity for the mission, and was the voice heard by anyone listening to the live feed from Mission Control. The Department of Defense (DOD) also had a console on the back row. This position supported recovery operations for astronauts after splashdown during programs before the shuttle, and was a key member during the various classified DOD shuttle missions.[101]

During Skylab, the FIDO and RETRO controllers generally worked in the control center for only a few hours each day. Each morning they updated the orbit calculations and ensured that the station was on the correct path. Typically, they did not need to work in the MOCR for the entire day, because Skylab was merely maintaining its orbit.[102]

Flight controllers generally worked nine-hour shifts, with an hour overlap on either end of the shift for updates about mission progress.[103] Typically each incoming flight controller reported one hour early to a conference room. There, one of the flight controllers from the on-duty team briefed those of the next shift for about fifteen minutes. At the same time, a member of the current team updated the incoming Staff Support Room team. After the flight controllers' briefing, each controller with an SSR reported to that team and updated them on the upcoming shift. Then, all flight controllers reported to Mission Control, where each controller was further briefed by the outgoing controller at their particular console.[104] This procedure, including redundancies that outsiders could see as unnecessarily wasting time, ensured that all pertinent information was passed from one team to the next and attempted to avoid having any surprises.

The extended shifts during missions created a distinct atmosphere in Mission Control. Gene Kranz talked about a particular smell generated by the controllers working in the room for hours. It was like stale pizza and sandwiches, burnt coffee, full wastebaskets, and the energy of anticipation that pervaded the atmosphere. That energy also led to a buzz of conversation, highlighted by brief dialogues of quick sentences using Mission Control jargon.[105]

With the original system, JSC had hardwired the command buttons and event lights on each console to specific processes. If an event light turned on, the controller pushed any of a number of command buttons to deal with the problem. Unfortunately, the console was so inflexible that if a change were needed with the buttons, it took months to reroute the sometimes thousands of wires.[106] A controller could not simply print out a screen image if he or she needed the information. Instead, the controller pushed a button with the appropriate command. Another console, with a thirty-five-millimeter camera, took a picture of the screen and then printed out a paper containing that image. The controller then received the paper with the image through the pneumatic tube system that connected the MOCRs to various other areas of the MCC.[107] These provide excellent examples of the limitations placed on the users by the system used to control the spacecraft.

Following the Apollo program, JSC deactivated the third-floor MOCR, and the second floor controlled all Skylab missions and the ASTP. In 1979, JSC reactivated the third floor for the space shuttle. Both MOCRs went through a major upgrade. NASA replaced the old console systems with a new console interface system (CONIS). At the same time, NASA updated the second-floor consoles by removing and repainting them tan, to coordinate with the new tan carpeting and walls (the third-floor MOCR has green consoles, hence the names "Brown MOCR" and "Green MOCR"). This provided a greater distinction between the two MOCRs.[108]

While Mission Control itself remained relatively constant throughout its history, NASA made needed upgrades as technology progressed. As JSC moved further into space shuttle missions, it recognized some problems with the MOCR. Most notably, the infrastructure was relatively inflexible and it was growing obsolete. Also, veteran flight controllers, who came of age on the consoles, were steadily replaced by new flight controllers with more computer experience. These concerns fueled the move for change and influenced the construction of the new Flight Control Rooms (FCRs).[109] By updating with off-the-shelf equipment, NASA has reportedly saved up to $30 million per year.[110]

Following the Challenger disaster on 28 January 1986, JSC once again evaluated the MCC system. More upgrades were implemented, including replacing the IBM 370/168 mainframes with four IBM 3083JXs, installed between January 1986 and September 1986. Due to its extensive use of the MOCR for classified shuttle flights, the DOD agreed to pay for a portion of the upgrade.[111]

Before the first desktop computers were officially installed in MOCR, some controllers had their own personal computers near their consoles. They wrote custom software for their needs, input information, and had the computer calculate solutions for them. While personal computers technically were not allowed by JSC, flight controllers became so adept at this process that they clamored even more for increased flexibility in the Mission Control hardware and software.[112] Interestingly, this seems to contradict previously mentioned concepts of control systems. It has been argued that control systems manage not only the object but the user, meaning that humans are subject to staying within the constraints of the established system. In this case, the flight controllers asserted their agency and changed the system to fit their needs.

The third-floor MOCR remained online for almost five years as JSC continued to transfer code from the old software. Finally, in October 2002, it was officially unplugged, decommissioned, and relegated to historical status. It remains a popular stop on the NASA Tram Tours of JSC.[113] The MOCRs had been designated a national historic landmark on 24 December 1985, twenty years after they first controlled *Gemini 4*.[114] This ensures that they will remain intact regardless of changes to the rest of JSC.

JSC realized in the mid-1990s that the old MOCRs could no longer handle space shuttle operations. They also were not acceptable for

White FCR, 2011.

control of the planned space station. JSC constructed three flight control rooms in the MCC. The White FCR operated space shuttle missions, the Blue FCR controlled the International Space Station, and a third room, the Red FCR, housed simulations to train new flight controllers.

The Red FCR came online first, in order to prepare controllers for the new rooms. The construction of the rooms, completed by BRSP, Inc., was relatively quick, taking only about six months for the Red FCR. This rapid construction was largely due to using infrastructure already in place, so there was no need to construct a new building.[115]

The White FCR cost approximately $250 million to complete.[116] It began operations with STS-70 (Space Transportation System) in July 1995, a mission that inserted one of the tracking and data relay satellites (TDRS) into orbit that proved vital to the communications network. The original MOCR continued to handle certain critical aspects of missions, including launch and landing, for another two years as the new FCR was phased in.

White FCR view of screens, 2011.

The White FCR had consoles in five rows with a somewhat irregular pattern, though the location of consoles remained similar to the previous rooms.' The first row was still called "the trench" and included the trajectory, FIDO, guidance, and GC departments. The second row consisted of systems controllers, such as propulsions, GNC, MMACS, and EGIL. The third row included more systems controllers, like the data processing and system engineer (DPS, pronounced "dips") for the computer systems, the assembly and checkout officer (ACO) in charge of payloads, FAO, and EECOM. The fourth row included INCO, the flight director, capcom, and PDRS for robot arm operations. Finally, the back row continued to host the PAO, MOD, booster or an EVA controller, depending on the aspect of the flight, and the flight surgeon.

With the International Space Station (ISS) continually manned since 1998, the Blue FCR had likewise had continuous controller presence. The Blue FCR had five rows of three consoles, with an additional console in the back right corner for the public affairs officer (PAO). Each row included one console on the right (from the front of the room) and two on the left, with a wide walkway down the middle. The flight director and Capcom occupied the fourth row from the front. The majority of other positions were systems positions and special positions created to operate the ISS. Some had similar tasks as those for the shuttle, but with slightly different names and call signs.

A typical ISS flight control team consisted of twelve to fifteen controllers, including the flight director. JSC designated most flight controllers and flight directors to work either the shuttle or ISS. When the shuttle docked with the ISS, the flight director and controllers in the Blue FCR took control of virtually all operations. Thus, the ISS flight director had overall control for flight operations, and the shuttle FD deferred to him or her. An ACO served as a liaison between the two FCRs.[117]

Due to a lack of space in the MCC, the Blue FCR was significantly smaller than the other Mission Control rooms. This problem eventually led JSC to move ISS operations to the second-floor MOCR, now called FCR 1, in 2006. JSC updated FCR 1 with new blue consoles set in four rows on the same level, losing the tiered approach of the MOCRs. Looking from the front of the room, the right side included either one

Blue FCR, 2011.

FCR 1, 2011.

(front row) or two separate consoles. The left side had either three or four consoles set in long rows. This setup seemed somewhat lopsided to the visitor. The flight director and capcom were located next to each other in the third row. Since the changeover, JSC utilized the Blue FCR only during special missions, such as STS-125, the final mission to the Hubble Space Telescope in May 2009, and STS-134, which installed the Alpha Magnetic Spectrometer (AMS-02) on the ISS in May 2011.

One important change with the new control rooms involved the basic level of adaptation. In the MOCR, the hardware required changing for upgrades. In the FCRs, personnel made most needed changes with the software, a much easier and quicker fix.[118] The FCRs also used off-the-shelf front-projection screens, which cost less than a single projector bulb for the old system, or about $75,000.[119]

JSC began a Mission Control Center Workstation, Server and Operating System Replacement (MWSOR) project in 2003, which replaced old hardware and software. With greater flexibility and more off-the-shelf aspects in the system, it decreased costs and greatly aided the overall system. JSC also began to use MCCx, a computer system that allowed flight controllers to log on and work from their home or office using a personal computer.[120] Lockheed Martin holds the current contract to operate and support the FCRs. The facilities development and operations contract (FDOC) began on 1 January 2009 and ran through 30 September 2012. In December 2011 NASA extended the contract for another year, bringing the total value to $919.5 million.[121]

The essentials of Mission Control remained relatively constant throughout its history. Controllers planned for a mission, trained for a mission, and then executed the mission. The planning and training included such detail that the controllers foresaw and handled virtually any anomaly during the actual flight of the mission. The more current flight controllers could sense continuity with the past. That these essentials remained intact indicates that the early Mercury and Gemini flight-control pioneers knew from the beginning how best to fly a mission.[122]

LAUNCH CONTROL COMPLEX

Houston is not the only Mission Control Center for human spaceflight in the United States. All missions are launched out of the Kennedy Space Center in Cape Canaveral, Florida. KSC houses a Launch Control Complex (LCC) that monitors spacecraft from their time on the launchpad until they clear the tower. Thus it is important to briefly discuss this vital link for JSC.

The history of KSC and the reason for its location in Florida has been discussed elsewhere. In essence, NASA took over the army's launch complex at Cape Canaveral and has maintained a presence there ever since. The first Launch Control Room was added in 1957 and consisted of just enough equipment to monitor test rockets.[123]

The first Mission Control for human spaceflight, Mercury Control, was built at the Cape, largely so that the controllers could have a direct line of sight, using periscopes, or by closed-circuit television, on the rockets as they launched. In case a problem occurred on the launchpad or immediately thereafter, the controllers could see the anomaly and abort the mission, if necessary. Control also needed to be located near the launch complex due to the constraints of the late 1950s and early

1960s technology. Signals were too weak to travel much farther without harmful degradation. As a result, NASA constructed blast-proof block-houses only a few hundred yards from the original launch area.[124]

While most of the flight controllers came from Langley, in Virginia, the early launch operations personnel commuted between Florida and Huntsville, Alabama. Most of these men were part of the German rocket team that formed the Marshall Space Flight Center (MSFC) and were led by Kurt Debus. As the launch operations wing of NASA solidi-fied at the Cape, the ties with the MSFC were split and they operated independently.

When the main Mission Control moved to Houston, NASA decided to maintain a Launch Control Complex in Florida. Launch Control is in charge of the spacecraft from its time on the launchpad until the rocket clears the launch tower, roughly ten seconds into flight. At that time, control is handed over to JSC. While Launch Control controls the spacecraft for a relatively short amount of time, it nevertheless plays an important part in the ground segment of American spaceflight.

Robotic spaceflight missions launched out of KSC work indepen-dently. The majority are launched from the Air Force facilities on Mer-ritt Island. Rather than a centralized Launch Control, they are operated out of blockhouses near each launchpad. Like the human spaceflight missions, control then reverts to the mission's main control room, whether it be at JPL, Goddard, or in a foreign nation, after the vehicle clears the tower.[125]

The Launch Control Complex (LCC) is an eight-story building with four levels adjacent to the massive Vehicle Assembly Building. Both are around three and a half miles from the launch site. Rather than rely on periscopes, the LCC includes large windows facing toward the launch area for direct visualization. The windows have screens with aluminum frames acting as blinds to reduce glare from the sometimes harsh Florida sun.[126]

Much of the checkout of rockets performed by the LCC has been au-tomated over the years. This helps to eliminate human error and allows for a smaller contingent of personnel in the room. NASA, however, in-sists that every contracting company that built the spacecraft must have a representative on hand in the LCC for quick reference. This mirrors the other control rooms. For instance, any Apollo mission with a lunar module on board had a Grumman representative (the contractor for the LM) in the MOCR.[127]

Automation only increased prior to the first shuttle launches, with improved electronics and computing. These advances even allowed the LCC to run substantial simulations with the launch hardware. As the LCC increasingly relied on computers, controllers were forced to adapt and change their focus from monitoring spacecraft to computer programming and debugging.[128]

The director of launch operations acts as the flight director of the LCC. He oversees all operations, and ensures that each controller and all systems are "go" for launch. This must be verified by the main Mission Control for that particular launch. The director of launch operations is assisted closely by the test conductor and the launch director. After all systems are checked out, the computer takes over, with eleven seconds remaining before launch, and can only be overridden by the humans in the LCC if the automation fails.[129]

The launch team's job is not finished after handover. They still must inspect the launch information and the ground systems. Any anomalies, no matter how seemingly slight, must be reviewed and fixed for future launches. While others are focused on the ongoing flight, the controllers and other workers at KSC have other priorities. They cannot relax until everything has been inspected. This is yet another example of the unheralded, sometimes unknown, and certainly unglamorous facts of spaceflight work.[130]

CONTROL ROOM WORK

There is more to Mission Control than just overseeing spacecraft. Like astronauts, controllers train for months before each mission flies. Just to get on console, controllers must endure extensive preparation, often years of it. They will spend somewhere between ten and one hundred times more time on console for training than for the actual mission. The majority of this training and preparation comes through simulations of missions. Some simulations involve Mission Control alone, while others include the astronauts in a separate building in their own simulators. Simulations are vital to understanding not only the missions but also the technology involved in Mission Control. All controllers must understand the optimal output for a mission, the constraints of the technology system, and any potential problems with the spacecraft or Mission Control.

Most missions will undergo a similar training regimen. Early simulations are meant to build a baseline for a mission. The simulators add few, if any, anomalies so that the controllers can prepare for how the mission should run. Slowly, the problems simulated increase in number and difficulty to test the controllers, the systems, and the procedures. The simulations are designed to push everyone involved to understand the systems and how to either work around or fix problems. In general, the simulators work with only one rule: there must be a viable solution to the problem. In other words, there are no *Kobayashi Maru* scenarios.[131]

Simulations for NASA's human spaceflight missions began before the move to Houston. Before the Mercury missions flew, NASA administration realized that controllers needed some training or simulation of a flight to prepare for the real thing. They built a temporary training area near Mercury Control, including separate rooms for each of the remote sites. Valuable lessons arose from those simulations, including the need for teamwork and discipline.[132] The simulations gradually became more sophisticated as technology improved, remaining an integral aspect of spaceflight success.

Chris Kraft stated that simulation was one of the most powerful tools developed during Mercury. It provided experience for everyone involved: flight controllers, astronauts, and communications network personnel.[133] Simulations have been described as the "heart and soul" of Mission Control.[134] Many valuable lessons came out of the simulations. In fact, the simulation procedures served as one way to assess potential flight controllers. Flight directors and other officials evaluated how the controllers handled the experience of simulations and determined whether or not they were fit to work missions. Controllers sometimes made life-or-death decisions in mere seconds, and the simulations provided necessary practice and determined whether or not they were ready to make such decisions.[135] In many ways, the simulations, along with the flight rules, were the keys to a successful mission, from the controllers' perspectives.[136] With more recent shuttle and ISS missions, the astronauts and flight controllers together complete between 80 and 115 hours of integrated simulations. This number has grown smaller over the years as the training and simulation process became more efficient.[137]

There is a story about a simulation that may be apocryphal but nevertheless helps to show the importance of simulations. A flight controller on the lower level wanted to talk to a flight controller on the level above him in the original MOCRs. When he stepped up, he used his hand to steady himself by grabbing on to the top of the console. As he did so, he accidentally pushed a number of command buttons that sent commands directly to the spacecraft. If that had happened during a mission, any number of problems could have ensued. Because it was a simulation, the controllers learned from an honest mistake. As a result, they installed covers over the command buttons to avoid inadvertent commands.[138] This is just one example of how simulations can bring about unexpected but necessary changes to the Mission Control environment.

Again, simulations prepare the controllers not only to run the missions but also to fix any problems that may materialize. Despite NASA's best attempt to avoid mission anomalies, inevitably some have occurred. Mission Control has served as a vital aspect of saving several missions.

The flight controllers of JSC must always remember that their decisions can mean life or death for the astronauts in space. If a problem arose during a flight, therefore, the controllers were trained to react carefully. In fact, they were told that unless they knew exactly what was happening and how to fix it, they were to do nothing. They were to verify the complete extent of the problem before they tried to fix it; otherwise they could make it worse. There was to be a measured balance between taking the time to recognize the entirety of the problem and to find a solution as quickly as possible. Many flight controllers point to *Apollo 13* as the prime example of this theory in action.[139]

The first three days of *Apollo 13*'s flight went by nearly perfectly. After a television broadcast on the third night, the electrical, environmental, and communications systems engineer (EECOM) requested a routine stir of the cryogenic tanks before the astronauts began their sleep cycle. The cryogenic tanks hold liquid oxygen and liquid hydrogen, which produce electricity, oxygen, and water, probably the three most important elements for human spaceflight. The gases settled into layers of different temperatures and densities, so it was vital to stir the tanks with small fans to prevent layering and allow for accurate readings.

The astronauts actually recognized that something had gone wrong before the controllers did. Across the control room, controllers began to see unusual readings, but most immediately presumed it was simply bad data or instrumentation problems and nothing serious. The flight director on console, Gene Kranz, did not know of any difficulties until the now famous call from astronaut James Lovell: "Houston, we've had a problem."[140] Kranz asked the flight controllers for more information, but few could provide any answers.

The next few minutes were punctuated by confusion and misinformation. Mission Control protocol held that controllers would not act without full knowledge of the best course of action, but reports coming in from the astronauts required some response. The immediate concern arose from one power distributor showing no power and the other continually dropping. As the EECOM talked with his support room, other flight controllers began to call those not at consoles to come to the MOCR for additional help.

One of the controllers in the spacecraft analysis room phoned John Aaron, who was at his home preparing for some much needed rest. As the SPAN controller told him the status and readings, Aaron immediately recognized that they had a real problem, not just faulty data. He rushed to the control center where he found it took some time to convince controllers they had a spacecraft problem, not an instrument problem.

Not only was *Apollo 13* losing power, but it was also losing oxygen. More than an hour after the accident began, flight controllers requested that the astronauts close the reactant valve for fuel cell 1 to stop the venting of oxygen. While the crisis had been placed on temporary hold, the command module did not have enough power or oxygen to keep the astronauts alive long enough to return home. Flight controllers recommended a desperate solution: use the lunar module as an emergency "lifeboat" for the crew's return trip.

The lunar module (LM) was designed to sustain two astronauts for up to forty-five hours. Instead, it had to keep three astronauts alive for between seventy-seven and one hundred hours.[141] Before transferring to the LM, the astronauts had to power down the command module in a matter of minutes, a procedure designed to take hours. After a successful transition, the controllers had to focus on how to bring the astronauts back safely to earth.

Gene Kranz, whose team had been on console during the first hour of the ordeal, took drastic action. He decided to take his team off console for the remainder of the mission, until reentry, so that they could focus on determining how best to overcome the many anomalies. Kranz placed John Aaron in charge of designing a procedure to power up the command module for reentry using the minimum amount of power left after the accident.

The controllers quickly agreed to use a free-return trajectory, allowing *Apollo 13* to continue to the moon and using lunar gravity to propel the spacecraft back to earth. In the meantime, controllers also devised a pump to scrub the carbon dioxide levels and preserve life-saving oxygen. The astronauts and controllers worked through a number of other smaller incidents, and three days after the accident, *Apollo 13* successfully returned to earth.

Apollo 13 has been called the "successful failure." The number of problems and failures overcome by NASA has never been equaled. In fact, the extent of the failures was so great that NASA acknowledged that they could not have simulated such an accident because it would have been dismissed as not realistic.[142] While *Apollo 13* is the most famous rescue mission, it was neither the first nor the last such episode.

When Neil Armstrong and Buzz Aldrin landed on the moon on *Apollo 11*, they cemented their place in history. Few people realize how close they came to never actually landing. As the astronauts began their descent, an alarm sounded with the designation 1202. The guidance support room recognized this as an indication of information overload. Essentially, the computer was receiving too much information and was attempting to start from the beginning of its computation list.

While this was not a normal reading, it was not so critical that the landing could not continue. The controllers told the astronauts to ignore the alarms and identify any subsequent alarms. Interestingly, the controllers knew of this situation almost solely because of a simulation with the same reading just fifteen days earlier.

Similar alarms registered for the remainder of the landing. Although the controllers could not be certain that nothing was wrong, because they did not know the cause of the alarm, the training and simulations told them that the best decision was to continue the landing. Without the fortuitous simulation two weeks prior, the controllers may not have known the nature of the alarm and might have unnecessarily aborted the landing.[143]

The *Apollo* 12 mission also involved a mission-saving decision by men in the MOCR. More than the others, this rescue was possible only thanks to a remarkable set of circumstances. Without a specific controller (John Aaron), who had earlier witnessed a specific anomaly during a simulation, and a specific astronaut in the correct seat (Alan Bean), it could have been doomed mere minutes into the flight. *Apollo* 12 remains a remarkable example of the necessity of not only training and skill but also luck in spaceflight.

America's first space station, Skylab, provided a series of other tests for the controllers on the ground in 1973 and 1974. Only a minute after the launch of the space station, the force of the launch ripped off the micrometeoroid heat shield.[144] Nine minutes later, one solar array broke off, while the other was not able to deploy fully due to an obstruction. The loss of the shield caused the station to heat up to dangerous levels, while the lack of fully deployed solar arrays limited power in the station. The launch of the first crew, originally scheduled for the following day, was postponed ten days until engineers across NASA could devise a solution for the two problems.

During those ten days, Mission Control endured some of its most trying times for the Skylab mission. They had to monitor the station constantly. In order to produce the most power, the auxiliary solar panels for the telescope needed to be pointed toward the sun, but this orientation also left the area of the station missing the heat shield exposed, and temperatures soared. The controllers then had to reorient the station in order to cool it down. These roll maneuvers had to be carefully timed to balance the power needed for the station and moderate the temperatures inside.

Once the astronauts finally began living onboard Skylab, the controllers settled into a routine. Small problems seemed to arise constantly to the extent that the men who worked on Skylab became known as "astronaut repairmen."[145] While the astronauts carried out each fix, controllers and other ground personnel had to devise the procedures.

The second crew, alone, had to repair the station's gyros, which maintain stability, replace the heated water dump probe, replace the tape recorders in the laboratory, remove circuits in a videotape recorder, and fix leaks in various systems.[146] One of the more amusing examples of repairing the station occurred during the first mission. While Commander Pete Conrad was outside the station for an EVA, Mission Control relayed a strange request. One of the batteries was not working properly, so in the age-old story of kicking or hitting something mechanical or electrical if it does not work, Conrad was told to tap the area with a hammer. When he did so, the battery sprang back to life.[147]

Despite NASA's makeup as a civil agency, and despite President Dwight D. Eisenhower's wish to keep it completely separate from the military, interactions abound. The early manned spaceflight programs relied on spacecraft that splashed down in the ocean, where the US Navy was the only organization with the infrastructure and experience to achieve recoveries. Thus, the Mission Control Center included a separate room for the Department of Defense recovery operations and the MOCRs had a console for the DOD. The DOD presence also further solidifies links between Mission Control and the Cold War.

Mission Control has had a strong military presence, due to the abundance of early controllers with military backgrounds. Some, like Gene Kranz, had flown in Korea as Air Force pilots. Many others had worked for the military as air traffic controllers, signal operators, communications experts, or missile trackers. This common military experience created a military-like air in the room and a dedication to discipline. Although later classes of controllers tended to have less military experience, the precedent remained largely intact.

Perhaps unknown to the general public, the military presence grew stronger during the shuttle years, with numerous classified DOD missions. Controllers had to pass a security clearance in order to work those missions. The majority of the DOD missions are still classified, so specific information available remains scant, though many are believed to revolve around spy satellites. Many of these satellites provided surveillance on the Soviet Union and the eastern bloc, reinforcing the central role of JSC in the Cold War.

Controllers in each of the centers must process vast amounts of information. Each station, however, is just one part of a much larger organization. Even if the controllers' data seem perfect, they may hint at anomalies in other systems. If the controllers are too focused on their own work, they may miss a critical point where their information could solve another controller's problem. Controllers must constantly communicate with each other in order to understand the bigger picture of the mission.

This vital aspect of Mission Control has already been illustrated with the *Apollo 13* rescue operation. If the EECOMs had focused solely on their own data, they may have persisted in thinking that it was only an instrumentation issue. By understanding that other controllers were experiencing faulty data, and by examining the larger picture of the mission, they were able to refocus on the true nature of the accident and work toward a solution.

Flight controllers in each of the control centers must be not only good engineers, but also good operators. As Gene Kranz explained, an engineer knows how a system works theoretically. An operator must know the theory, as well as have the knowledge and experience about how systems work together to accomplish a mission. Controllers must have knowledge well beyond just their individual system and responsibilities.[148] To gain this knowledge, however, JSC, like the other control centers, does not provide formal training. Instead, the controllers needed to teach themselves the technical aspects of their systems. Many times the various controllers working the same console worked together to gain that information. They could then create a more comprehensive knowledge base that was used to create the flight rules for each mission.[149]

Communicating within control rooms can be chaotic. JSC's Mission Control is especially famous for its nearly unintelligible voice loops, or voice communication systems. Controllers use headsets to plug in to different voice loops. For instance, if they need to talk with their support rooms they can simply switch over to their dedicated back-room voice loop. Most remain on an open flight director loop, where all conversations can be heard at once. At any one time there can be dozens of voices talking. To the novice, it can be disorienting, or even appear to be pandemonium. Controllers must quickly learn to drown out the majority of the voices as white noise while listening for keywords. If they cannot do so, they will not last long in Mission Control. A trained human ear remains a necessity in the seeming chaos of Mission Control, especially when technology cannot stem the flood of information.

Written communications could be as simple as passing a note to a nearby controller. The MOCRs of JSC utilized pneumatic tubes for more formal interactions. For instance, the old consoles did not include printers, so if they needed a printed image of a screen they pressed the appropriate button and it printed in a separate room. The printer operators then placed the document in the pneumatic tube and sent it to the proper controller.

Email and the Internet have changed information exchange in control rooms, much as they have in everyday life. Communication now is virtually instantaneous, or at least as instantaneous as the controller's access to email service. Some systems have even advanced to the point where controllers can access them online and complete their work from their home or office. Despite this capability, controllers continue to complete the majority of their work in Mission Control itself, where they can maintain personal interaction with their colleagues. Thus, while the new system does not invalidate the necessity of a control room, controllers will take advantage of any technology that makes their job easier.

During Gemini and Apollo, all communications between the Mission Control in Houston and astronauts in space were open communications. Anybody with the capability could listen to those conversations, which was most notably used by the press. Before Skylab, NASA agreed that, with an extended-duration mission, the need for private conversations might arise. There might be emergency situations that did not need to be shared with everyone, and certain communications between the astronauts and the flight surgeon might warrant doctor-patient confidentiality. Therefore, NASA added a private line for communications with Mission Control.[150] This idea was met with some reservation; however, it proved to be a wise change for future missions and has continued to the present.

*   *   *

Since it was built from scratch, NASA could construct the MCC specifically for controlling NASA's manned missions. Like most technological elements, it required numerous updates and changes over its decades of existence. JSC still strives to provide the best resources to continue the prominence of NASA's manned spaceflight.

Following the final shuttle launch in 2011, the human spaceflight program consisted of the International Space Station and vague promises of future flights from a proposed new rocket. Despite being the home of Mission Control and astronaut training, JSC employees, not surprisingly, worry about their future. If the United States commits to the proposed human spaceflight program, JSC is poised to remain at the heart of those efforts for decades to come. Until that time, Houston, rather than being the preeminent Mission Control Center, serves as merely one in a series of relatively equal space station control centers.

2

JET PROPULSION LABORATORY

During the 1960s, nearly everyone interested in spaceflight focused on the Apollo program. As a result, the Jet Propulsion Laboratory's robotic missions launched and flew in relative obscurity. These missions, however, made Apollo feasible. The Pioneer program proved that spacecraft could be launched within close proximity to another large object in space. More importantly, the Surveyor program successfully landed on the moon. Some laypeople worried about the composition of the lunar surface and whether or not it was safe for humans to land there. NASA, however, had working knowledge that there was little to no potential for a spacecraft to be swallowed by the moon.

Such is the fate of unmanned, robotic spaceflight. These spacecraft are largely ignored, but they are always on the vanguard, pushing humanity further into the undiscovered blackness of space. They were the first to rendezvous. They were the first to land on the moon. They have made the first steps to Mars, Venus, Mercury, and the outer planets, far before human spaceflight has legitimately prepared for such missions. In many ways, robotic missions deserve as much, if not more, recognition than the human spaceflight programs.

NASA includes many centers that focus on robotic spaceflight, but the primary location remains in Pasadena, California. The Jet Propulsion Laboratory (JPL) maintains the Main Control Room for unmanned missions. Developed independently at roughly the same time as the control center in Houston, the JPL control room serves as a vital model for evaluation against the Mission Control Rooms at JSC.

JET PROPULSION LABORATORY

The Jet Propulsion Laboratory in Pasadena, California, began as the interest of a few scientists and engineers at the California Institute of Technology (Caltech) and blossomed, over some decades, into one of the premier spaceflight control centers in the world.[1] The Guggenheim Aeronautical Laboratory at Caltech (GALCIT) focused on aerodynamic research. Until the mid-1930s, it dismissed rocketry as nonacademic. Theodore von Kármán, director of GALCIT, changed his mind on the relevance of rocketry research, thanks to a series of graduate students. One of them, Frank J. Malina, led a small group of experimenters, including John W. Parsons and Edward S. Forman, to the first rocket motor tests at GALCIT in 1936 and 1937.[2]

After a few years of testing, the Army Air Corps encouraged the National Academy of Sciences to present GALCIT with a grant to research military applications for jet propulsion in 1939.[3] Thus began a long relationship between the rocket scientists of Caltech and the military. With this funding, GALCIT completed the first successful jet-fuel-assisted takeoff (JATO) of an airplane in the United States on 12 August 1941 at March Field, in Riverside, California.[4] Just eight months later, they accomplished another major feat. At Muroc Army Air Field, later known as Edwards Air Force Base, a successful takeoff of a Douglas A-20A marked the first American plane with permanent rocket power.[5] GALCIT tested advancements with both solid and liquid propellants. Some scientists and engineers also worked on guidance and control of the missiles for more accurate deployment. Some aspects of the control included radar tracking and radio signals from the ground to correct the flight path.[6] Thus they created a primitive ground control system.

As GALCIT grew and its connection with the Army Air Corps (later Army Air Forces) strengthened, its officials investigated the possibility of improving its relationship with Caltech. Many of the academics at the university objected to any connections with the military. After some contentious negotiations, the Jet Propulsion Laboratory, still considered an aspect of GALCIT, began work on guided missiles on 1 July 1944. Although the initial site remained Army Air Forces property, the close connection with Caltech continued, because a large portion of its staff came from that highly acclaimed university.[7] Experiments continued with both jet engines and rockets. GALCIT made history on 11 October 1945, when a WAC Corporal rocket reached forty miles in altitude and became the first American rocket to escape the earth's atmosphere.[8]

Following World War II, the Army and Caltech debated their interest in continuing to support JPL. The Army viewed rocket and missile research as a top priority for future technologies as the country transitioned to the Cold War. Caltech officials had to decide between objecting to the military and reaping the benefits of government funding. On 1 April 1946, the Caltech board of trustees officially approved their continued relationship with JPL.[9] Thus, the Jet Propulsion Laboratory served as a significant example of the growing military-industrial-university complex in the United States during the Cold War.

The connection to the military, especially during the Cold War, brought unforeseen problems for Caltech staff. Mandatory regulations forced all scientists and engineers to undergo security screenings. The system scrutinized foreign nationals, especially those from Communist China. The government classified Dr. H. S. Tsien, for instance, both as a security risk and as an undesirable alien. This occurred despite his prominent role in the creation of JPL and its subsequent success. His security risk status outweighed his alien classification, which forced him to stay at Caltech but prevented his working on classified materials. He remained at Caltech until 1955, when the Immigration and Naturalization Service deported him to China, where he subsequently led the Chinese missile programs. Despite the security hurdles, the connection between JPL and Caltech became an important recruiting tool for the nascent laboratory. Some staff had opportunities to teach at the university, adding to the lab's attractiveness.[10]

Throughout the latter half of the 1940s and the 1950s, JPL continued to make strides in both solid- and liquid-fueled rockets for the Army. These included the Corporal and Sergeant series of missiles. As Cold War tensions grew and the military-industrial complex became more entrenched, the laboratory transitioned from almost pure research and development to include large-scale assembly of missiles. Indeed, the military applications became so intertwined that in 1953, the Army attempted to appoint an officer to command JPL. Caltech, not surprisingly, balked, and the Army rescinded the request. Regardless, the laboratory could no longer downplay its military connection.[11]

As the country moved toward a postwar mentality, some of the neighborhoods surrounding JPL complained about the presence of what was supposed to be a temporary installation used only for the duration of the war. The lack of permanent buildings, along with the drab paint and lack of landscaping, became an eyesore for many of the local residents. Many also objected to the noise from the rocket motor testing and flashing lights at odd hours. Some officials, including Caltech president Dr. Lee DuBridge, tried to shift the focus by arguing that Americans had to endure minor annoyances for the greater good as the country dealt with the real problem: the Soviet Union. In doing so DuBridge caused a greater rift to grow between the laboratory and the university, since his comments solidified the military applications of its work. Meanwhile, the two sides strove to find a more permanent solution.[12]

Tensions continued to mount as JPL's budget grew to more than twice that of Caltech in 1957, leading many at both sites to question the nature of their relationship. Some Caltech trustees viewed JPL as a possible hindrance to their educational mission, but they could not overlook the vast amounts of money the university received from the government via the laboratory. Ultimately they avoided these issues by agreeing to continue as they were in order to complete their projects in a timely manner.[13] The JPL-Caltech relationship remained contentious but not fully addressed for a few more years.

Shortly after the Soviet Union launched *Sputnik* in 1957, the United States began to work toward its own satellite launch. After some setbacks, the focus turned to *Juno I*, a rocket built by Wernher von Braun and the Redstone Arsenal in Huntsville, Alabama. Although the rocket was completed, it still lacked a proper satellite. JPL had already worked with the Redstone Arsenal on other projects, so naturally it expressed interest in this venture. After much deliberation, NASA awarded JPL the contract to build an American satellite to counter Soviet efforts.[14] JPL engineers worked around the clock, and by 31 January 1958, *Explorer 1* was ready for launch. JPL utilized its limited tracking abilities, including an early Mission Control Room and a few remote antennas, to monitor the spacecraft.

Next JPL officials had to make an important decision about the direction of their space program. They could remain closely tied to the Army and work for a military space program, or they could side with the newly formed civilian space program, NASA. On the military side, the Army slowly downgraded its space efforts and the Air Force expanded its role. On the civilian side, there were some hints that a nonmilitary venture was doomed to fail. To that end, JPL Director William H. Pickering even argued to the Eisenhower administration that NASA would fail without JPL as a centerpiece. After a series of negotiations, the Department of Defense agreed to allow JPL to join NASA beginning in 1959, but not without completing its work on existing programs, the most important of which was the Sergeant series of missiles.[15]

On 15 October 1958, only two weeks after the creation of NASA, the new space agency proposed to integrate JPL into the space program. President Eisenhower's Executive Order 10793, signed on 3 December of the same year, officially transferred JPL from the United States Army to NASA.[16] On 1 January 1959, JPL officially became a part of America's civilian space program. Unfortunately, the laboratory's ambition far outstripped its ability. JPL officials envisioned a JPL-dominated NASA, with long-range plans for reaching the planets beginning within a relatively short amount of time. NASA officials, on the other hand, stressed a more subordinate role for JPL, including research and technical advice, while the space agency focused on human spaceflight out of Langley and Huntsville. NASA and JPL managers took some time to agree on a middle ground.[17]

JPL experienced difficult growing pains in the first five to ten years of its existence as a member of NASA. For instance, NASA conducted business differently from what JPL had been accustomed to under the Army. It had a more restricted budget, which was monitored more closely. NASA also wanted more everyday management by Pickering himself, rather than the university-like approach. He should have oversight of nearly all aspects of the organization but generally stay out of day-to-day decisions.

There were also some issues that developed with the relationship among JPL, Caltech, and NASA. JPL was a NASA center, with federal funding, but Caltech still operated it. Any major changes, or any potential programs, required the approval of both Caltech and NASA. Some employees had significant difficulties with this two-pronged leadership. After some time, JPL grew more comfortable with its role as one of many centers in the national space agency.[18]

The JPL expanded substantially during the mid-1960s into the campus-like facility it is today. Because the ability to expand horizontally was limited, due to geographic and residential reasons, the center grew vertically. Beginning in 1962, JPL built Building 180, which housed the administration and some engineering. They added a landscaped central area just south of 180. Farther south, JPL built the Von Kármán Auditorium, which abutted the main entrance to the laboratory.[19] JPL opened the Space Flight Operations Facility (SFOF), the home of Mission Control, the same day as the auditorium.

Like most of the NASA centers, JPL had to prove its worth constantly in the late 1960s and 1970s as the nation's attention turned to other issues, like the conflict in Vietnam and the Civil Rights movement. As NASA's budget shrank, JPL, like many of the other centers, had to work with less and vary its missions in order to remain relevant. JPL also was compelled to diversify its workforce. While roughly 30 percent of employees were either women or minorities, only a handful of them served in managerial positions. Perhaps one may attribute these problems to the lack of women and minority engineers in the workforce in the 1970s, but NASA also deserves some of the blame for not encouraging more aggressive recruiting.[20]

As the budget continued to decline in the 1980s, particularly after the end of the Cold War and its connection with technology, JPL adopted the idea of "faster, better, cheaper" under NASA administrator Daniel Goldin. With this new approach, JPL sought less ambitious missions that still provided valuable scientific outcomes. Unfortunately, under this model, JPL experienced a few failed missions among the successes. *Mars Observer*, *Mars Climate Orbiter*, and *Mars Polar Lander*, all in the

Aerial view of JPL, 1961. Courtesy of NASA, http://grin.hq.nasa.gov/ABSTRACTS/
GPN-2000-001980.html.

1990s, enhanced the red planet's reputation as a spacecraft destroyer, but in reality their failures can generally be explained by simple yet costly mistakes. Some within the spaceflight community argue that JPL embraced "faster" and "cheaper" while overlooking "better."

After these failures, JPL moved away from the "faster, better, cheaper" concept, focusing instead on medium-sized projects that promised more output without some of the excessive budgets of the larger programs. It has also embraced missions with international partners to help reduce costs. JPL continues to adjust to the times and remains at the forefront of interplanetary spaceflight.

SPACE FLIGHT OPERATIONS FACILITY

The Jet Propulsion Laboratory included only a temporary control area to monitor missile tests in the 1950s. When JPL joined NASA, the existing infrastructure simply could not suffice for future programs. If JPL were to claim its place as NASA's primary robotic spaceflight facility, it would require a more permanent control center.

JPL's Data Handling Committee wrote a report on 22 June 1961 stating that the Communications Center could not manage future Ranger and Mariner missions. The interim report suggested that JPL upgrade its data processing, build a new facility to handle the data and flight operations, and construct the facility as soon as possible before the existing center became obsolete.[21] Construction began almost immediately.

Before designing and building the Space Flight Operations Facility (SFOF), designers and engineers relied on experience from earlier missions. First, they recognized the need for a centralized control area. They also realized the need to make some critical decisions in a short amount of time. Finally, they had learned some of the operating requirements for long-duration missions.[22] There is no indication that they acquired information on how to build a control room from outside sources. These lessons played key roles in various aspects of SFOF.

JPL officially dedicated the Space Flight Operations Facility, as well as the Central Engineering Building, the Space Sciences Building, and the Von Kármán Auditorium, on 14 May 1964.[23] The SFOF was opened by receiving a signal bounced off Venus, a total of eighty-three million miles and taking seven minutes, twenty-five seconds.[24] During the dedication, Homer E. Newell, the associate administrator for Space Science and Applications for NASA, remarked that its roles included collecting data from the Deep Space Network (DSN), reviewing and analyzing those data, translating data into commands for each mission, and centralizing command of missions. It necessitated constant surveillance during missions. Marshall Johnson, chief of the Space Flight Operations Section, played a leading role in its planning and construction.[25]

The SFOF currently consists of buildings 230 and 264. Building 230 includes the Main Operations Room as well as various other supporting control rooms and offices. Building 264, originally built as offices for missions, also now incorporates additional supporting control rooms. Both are centrally located on the JPL campus, closer to the west main gate. Building 230 was directly north of building 180, which was originally built as the Central Engineering Building but now serves as the headquarters or administration building. Building 264 stands east and south of buildings 230 and 180. The campus cafeteria lies between and to the south of buildings 180 and 264.

JPL originally designed the basement of the SFOF primarily to house equipment to maintain the rest of the building, such as air-conditioning units, power, an emergency power system, and water heaters. The basement also held communications rooms, comprised of communications terminals, teletype, and communications control, as well as various telemetry data and processing areas. The first floor included a lobby, mission control and operations, and various other support rooms and analysis areas. Office space and other control rooms made up the second floor.[26] Upon its completion, SFOF had about fifty-five thousand square feet of operational space, with a total area of more than one hundred twenty thousand square feet.[27]

There were to be three main technical support areas for the controllers in the main room. The Space Science Analysis Room evaluated the data from the experiments. The Flight Path Analysis Room handled tracking and control commands. The Spacecraft Performance Analysis Room monitored the condition of the spacecraft. The SFOF also included a Planetary Operations Room during the interplanetary time of flight when there was no need for a full complement of controllers.[28]

The general layout for SFOF has remained relatively similar, though specific uses for various support rooms have changed multiple times throughout its history.

The Operations Room of the SFOF first controlled the *Ranger 7* mission in July 1964. The mission was so historic for the center that each division of JPL received specific scheduled times when they could observe the mission from the visitors' gallery, which has a capacity of about forty people.[29] The gallery remains the primary location for public viewings of JPL's Operations Room.

Glenn Lairmore began serving as SFOF manager in February 1964. His duties included the SFOF budget, contractor management, facility development and operations, and maintenance of the Operations and Development Procedures Manual.[30] On 1 December 1964, management responsibility for SFOF transferred from the Office of Space Science Applications (OSSA) to the Office of Tracking and Data Acquisition (OTDA).[31] With this change, SFOF began full operation.

In 1965, costs independent of those to run missions for SFOF were approximately eight million dollars.[32] That same year, the SFOF first supported multiple missions simultaneously with *Ranger 8*, *Ranger 9*, and *Mariner 4*.[33] In order to handle the massive input of communications, JPL installed an electronic communications processor, which greatly facilitated the transmission of communications.[34] New and better technologies helped the controllers with greater responsibilities to avoid the potential tidal wave of information from the spacecraft.

In the early years of SFOF, JPL recorded the majority of data from spacecraft on magnetic tapes at the various DSN stations before transferring it to SFOF. Especially vital information, however, could be transmitted immediately to SFOF and placed on magnetic tapes at JPL. Scientists and engineers had push-button control of information displayed for them in the control rooms. The space flight operations director had authority over mission control itself, while the SFOF manager watched over the whole facility.[35] For immediate communications between SFOF and the network stations, the center primarily used teletype, though voice communications could be used in emergencies.[36]

JPL separated mission-dependent and mission-independent facilities for particular missions or for all missions, respectively. Mission-dependent facilities only operated for a specific mission. Independent facilities could be used by any mission. For example, the control rooms in the operations area, support areas, facilities, and communications of SFOF were all independent. Technical and control rooms in various parts of SFOF were dependent.[37]

The majority of SFOF supported the Main Operations Room. In the technical areas, three teams backed up various control personnel while they were at their consoles. Space Science Analysis evaluated data from scientific experiments and issued commands. Flight Path Analysis evaluated tracking data and subsequent commands. Spacecraft Performance Analysis reviewed the condition of the spacecraft and issued pertinent commands.[38] These areas operated somewhat similarly to JSC's Staff Support Rooms, though with more direct operational control.

The Communications Center of SFOF handled both internal and external communications, along with the DSN. This room, located in the basement of SFOF, had a Communications Center coordinator who managed the routing of data and the use of internal and external communications facilities and equipment. A Central Computer Complex (CCC) on the second floor included two computer systems for primary and backup computations. The CCC coordinator handled the overall operational management of the data processing systems. A Planetary Operations Room, also on the second floor, served as a secondary Mission Control during interplanetary sequences when operations required less personnel.[39]

SFOF also included an Uninterruptible Power System (UPS), which provided emergency power. This system consisted of a series of diesel generators in the basement of SFOF. The diesel fuel was regularly replaced so as to avoid any problems with it.[40] Like JSC, JPL recognized the need for backups in the case of an emergency.

In all, SFOF, as originally designed, required fifty tons of wiring and cabling. The control rooms included 31 consoles, 100 closed-circuit television (CCTV) cameras, and more than 200 television displays. Each console had the ability to select 150 contacts and included a headset, telephone, intercom, and television. Digital displays could show up to 3,500 numbers. They had the ability to accept, process, and display

4,500 bits of real-time data, and up to 100,000 bits per second could be recorded for later use.[41] At the time, JPL utilized state-of-the-art equipment, but all of it necessarily underwent numerous updates throughout the years as more and better technology appeared.

During the Viking project, three computer centers supported operations. The Mission Control Computing Facility (MCCF) included three IBM 360/75 computers. This system had a one-megabyte main core with two megabytes of large core storage and 460 megabytes of disc storage. The General Purpose Computing Facility (GPCF) housed two UNIVAC 1108s. Finally, the Mission and Test Computing Facility (MTCF) held numerous UNIVAC 1219s, 1230s, and 1616s.[42]

Beginning with Magellan, the SFOF used UNIX software with a JPL-specific overlay, supplemented with Sun Microsystems servers. This software easily supported multiple missions. It also provided homogeneity across projects; previously each project had used its own system. That being said, some of the older missions still use their original software or hardware, such as punch cards for Galileo.[43] Thus, JPL's computing complex resembles a "strange space" of fading in/fading out technology, as discussed by André Jansson and Amanda Lagerkvist. The best and newest computer hardware is interspersed with decades-old technology still linked to older missions.[44]

While the MCC at JSC handled only one or, rarely, two subsequent missions, the SFOF regularly managed multiple missions. Due to the sheer number of missions controlled at JPL, mission operations quickly outgrew building 230. Building 264 was constructed in 1970–71 as the SFOF Systems Development Laboratory. While the original building had only two stories, plans called for six more in the near future. The two floors included thirty thousand square feet of workspace, with the potential for a total of one hundred twenty thousand square feet. From the beginning its objective was to serve as the home for mission support facilities and as a natural extension of SFOF as JPL and its missions grew in size, scope, and numbers.[45]

In the early 1980s, JPL planned a much-needed upgrade to SFOF. One of the planners' biggest concerns was a limited budget. Part of the overhaul included replacing the last remaining IBM 360/75, which they estimated cost at least three hundred thousand dollars.[46] The overall construction project, which ran from February 1982 to September 1984, had an early budget estimate of between $1.2 and $1.4 million.[47]

One of the most important differences between manned spaceflight missions at JSC and robotic missions at JPL is the distance they travel from the earth. As missions traveled farther and farther away from earth, the delay in communications increased. As a result, controllers anticipated the next command for a spacecraft, sometimes many hours in advance. For instance, by 1999, a round-trip communication between *Voyager* and JPL and then back to *Voyager* took more than twenty hours, which contrasts strikingly with the near-instantaneous communications with the International Space Station and other near-earth objects, or the roughly three-second delay to the moon.[48]

On 25 July 1994, the National Register of Historic Places named SFOF a national historic landmark, in recognition of its importance to the space program. The original form had been submitted in 1984 and stressed the building's importance to space exploration as an extension of earlier explorers like Christopher Columbus and Samuel de Champlain. It especially highlighted the Deep Space Network control center. The space simulator in building 150 and the twenty-six-meter (eighty-five-foot) antenna in Goldstone, California, were also made national historic landmarks the same day.[49] The designation as national historic landmark secures the SFOF its place in history. The control rooms remain highly active with the many continuing and planned future missions.

JPL OPERATIONS ROOM

The first control room at the Jet Propulsion Laboratory began operations in January 1956. It consisted largely of a terminal, office furniture, and some calculation equipment for data processing and orbit computations.[50] The control room also had a wall map for tracking satellites. This original control center had no digital elements.[51] It was enough, however, to control the early missions through the Ranger project. The control room utilized an IBM 704 computer for processing orbit data.

By 1961, JPL's control room for Ranger was located with the Pasadena Communications Center, which handled incoming data from the communications network. Although the two rooms were located next to each other to expedite the exchange of information, a glass wall separated them, making direct interaction nearly impossible. The center included an IBM 709 computer to process incoming data.[52] The computer, which used vacuum tubes, generated so much heat that it depended on an extensive air-conditioning system. The IBM 709 lasted two years, until JPL updated it with an IBM 7090 in 1962. The next year, JPL decided to pair two computers to process the influx of information better. They installed an IBM 7040 and 7094 with a 1301 disk storage file. There were some difficulties with the IBM 7040, so JPL quickly upgraded to an IBM 7044. Before the *Ranger 5* mission, JPL updated the Communications Center to include push-button switching, to make data exchange more efficient.[53]

After the SFOF took over control duties, the IBM 7044 and IBM 7094 played integral roles during the Mariner missions for tracking and data processing.[54] The IBM 7094 experienced hardware and software problems during the *Surveyor 1* mission, necessitating repairs to the card decks and card readers. There were also malfunctions in the interface between the 7044 and 7094. This issue became so severe that JPL had to involve IBM in a repair that took two and a half hours to complete.[55]

The operations management plan for the Mission Control and Computing Center (MCCC), prepared for the Viking missions of the 1970s, describes a typical setup for operations of a JPL mission. The MCCC operations control team (MOCT) handled real-time flight support as well as various aspects of control, records, and analysis. The MCCC operations control chief (MOCC) directed the MOCT, including scheduling, directing action in the case of an anomaly, and serving as a primary interface for operations. The computer operations chief (COMPUTER CHIEF) was responsible for computer support and operations and interactions with DSN. The communications chief (COMM CHIEF) managed internal communications for SFOF. The facility support chief (SUPPORT CHIEF) was in charge of the general facility, including power, janitorial staff, displays, and safety. An operations analysis chief monitored the operational performance and directed the MCCC operational analysis team (MOAT) during the evaluation of the systems and data. The data processing controller (DATA CHIEF) coordinated data processing in the Mission Support Area (MSA) as well as computer systems. The MSA also housed the analysis program operations personnel and the data gathering and distribution personnel. An MCCC facility and operations project engineer (FOPE) served as an interface for all the various elements of operations in the MCCC to make sure that JPL completed all requirements for the project. Finally, the MCCC operations manager supported the program from an overall JPL viewpoint, keeping in mind other programs and systems.[56]

An activity engineer (ACE) flight controller served as a type of capcom for JPL missions. This controller took any action necessary to successfully complete the mission, so far as those actions fell within the guidelines established for the mission.[57] Any decisions based on anomalies advanced through the proper channels before a command was sent.

By design, the room nearly always had low lighting, much like JSC's control rooms, allowing the controllers to view their screens better. It also aided with heat reduction, because the massive amount of electronics and hardware in a confined space could potentially raise temperatures, requiring increased air conditioning. Due to the low lighting, the Operations Room often was referred to as "the darkroom."[58]

Records for individual renovations at JPL are sparse, perhaps nonexistent. By viewing images of the Operations Room over the decades, however, one quickly realizes that the layout of consoles and computers changed frequently. In this way, JPL contrasts strikingly with JSC and ESOC, where the control room layouts largely did not change once constructed. The reason for this difference remains mostly speculative, though it may arise from the changes in missions and the variety of missions over the decades.

Beginning in 1964, controllers had individual workstations in straight rows. The open space of the Operations Room was significantly smaller, with only two or three rows of consoles. The front of the room did have the screens displaying information above glassed-in rooms for other controllers. The consoles were mostly gray-blue in color. Behind the main control area was another glassed-in area for additional controllers. The majority of consoles had two screens for displaying information or images, a telephone receiver, and push buttons with no keyboard. Ashtrays were another must. The room also included various analysis and operations areas. Glass panes separated these areas, which JPL stated created a unified atmosphere.[59] The view screens, located at the front of the room, came from the 1964 Republican Convention, which was held in San Francisco earlier that year.[60]

By 1968, during the Surveyor program, JPL replaced some of the consoles with round-table consoles and one long table in the center of the room. Some of the original consoles faced inward around the edges of the room. The new tables were largely brown and white, though the consoles themselves remained gray-blue. They covered the glassed-in rooms at the front with dark red curtains, giving the room a somber, reddish hue. Photos suggest that blue carpeting overlaid the old tile floors. They began using the round consoles, which could hold six workstations, with the Pioneer VIII, Surveyor, and Mariner missions. For Mariner, the round consoles included the communications controller, tracking chiefs, and multiple tracking stations for both Mariner and Pioneer. The long table held three stations in the rounded portion at the end of the table, which accommodated the Deep Space Instrumentation Facility (DSIF) operations manager, DSN operations manager, and the SFOF Ground Communications Facility operations manager.[61]

By 1981, after the *Voyager* launches, JPL replaced the long table with a half-circle console holding five screens in the center of the room to join the other round consoles. New tan carpeting was installed. They also exchanged the curtains at the front of the room with a black wall, making the "darkroom" even darker. Just two years later, the round tables were gone, and only five semicircular, five-monitor consoles remained. These consoles were yellow, and, with the tan floor, made the room nearly all yellow-tan and black. JPL has since given the round consoles to the Russians, who used them in their control center.[62] By 1986, JPL added at least three more of the rounded consoles, dramatically changing the amount of open space in the room.

In 1990, shortly after the *Magellan* and *Galileo* launches, at least one of the semicircular consoles remained, while JPL replaced the rest with more traditional, independent, and straight consoles. By 1995, following *Mars Observer* but before *Mars Global Surveyor*, all of the consoles were individual workstations, much like the original setup, and again the room had taken on a blue tint. They painted the walls white, making the "darkroom" significantly brighter than it had ever been before. There were more changes the next year, when updated consoles lined the front of the room, and individual computer monitors on large tables inhabited the center of the room. Before the launches of *Spirit* and *Opportunity* in 2003, the long line of consoles at the front remained. JPL placed the computer monitors in the center in a rounded layout closer to the current system. Some of the monitors were, by this time, flat-screens as well.[63]

JPL redesigned and renovated the Operations Room most recently in 2008. The SFOF managers met with designer Blaine Baggett to plan the aesthetics of the room for optimal working conditions. The current setup includes three rows of curved consoles. Each row generally has one controller overseeing each half of the computers. The curve allows the controllers to view all of their monitors and allows the rows to fit into the existing space. These consoles, designed and built by Evans Consoles, are some of the few consoles that are seismically rated, a particular concern for JPL given its location, proximate to major fault lines.[64] The room is significantly darker now as well, making the "darkroom" moniker more than apt. JPL also replaced the overhead lighting with individual lights on the consoles themselves.

The room is laid out analogous to the data flow. Information from spacecraft enters through the DSN control, the Network Operations Control Center (NOCC), which occupies the front of the room. It is then transferred to controllers, who reroute it to its final destination, at the program's specific control area. This data systems operations team processes, catalogues, and distributes the information. It consists of only two controllers, each of whom oversees half of the computer monitors on the row. Commands are then sent from those programs through the controllers back to the front of the room and the DSN

Operations room, 2010.

before transmission to the individual spacecraft.[65] In this way, any visitor with knowledge of the system can visualize the flow of information through the room quite easily.

The Operations Room can be described as a "throughput" facility. It only processes data, leaving the analysis of information to the individual program control rooms. In essence, it is more concerned with the quality and quantity of information processed than the actual content.[66] This is another example of a difference from the JSC control rooms, which process as well as analyze data.

The screens at the front of the room continue to provide some information about current missions, including which spacecraft are transmitting data at what times. The programs are color coded, and the scrolling of data represents transmission. One of the screens frequently shows a slideshow of images of the Operations Room through the ages. The managers admit, however, that they are on display almost exclusively for the benefit of visitors, similar to the screens at JSC.[67]

The Critical Events Mission Support Area is connected to the side of the Operations Room. This area includes controllers and support personnel for programs during critical aspects of missions, including launch, rendezvous, and planetary landings. This is also an area for VIPs, and is usually the area filmed for news reports of missions.[68]

A glassed-in area is located under the screens at the front of the room. This area includes the DSN controllers. The design reverts back to the original layout. Over the years the glass had been covered by curtains, and then replaced with a wall. With the renovation in 2008, the designers returned to the glass to give the room a more open feeling and to more logically connect the DSN controllers with the rest of the operations controllers. The DSN control employs contractors, which are currently supplied by ITT. An operations chief serves as a supervisor for DSN control, with a tracking support specialist assisting.[69]

JPL missions operate under the precept that "whoever builds, operates." This means that the company or contractor that builds the spacecraft is the one that has the authority to control the mission as they see fit. They operate from an individual Mission Support Area (MSA), which serves as the control room for the majority of the mission. These rooms have also been called Project Operations Control Centers (POCC).[70] Only critical aspects of missions are controlled from the Operations Room. Many Mission Support Areas are located within the two SFOF buildings; however, this is not a requirement. Companies may house their MSA anywhere they wish, including as far away as Denver, Colorado, for Lockheed Martin, and Greenbelt, Maryland, for the Goddard Space Flight Center.[71]

Similar to those at JSC, JPL controllers can be either employees or contractors. The ratio of employees to contractors varies greatly, however. Whether or not one is an employee or contractor is largely a numbers game. In fact, many switch between the two, depending on the current needs of JPL or the contractor. All activity engineers, in particular, must be employees.

The mission controller for each mission serves as the direct interface with the DSN for the processing of data. Mission managers or flight directors handle the day-to-day running of missions. Most control rooms work with three to four shifts of one or two people, depending on the project and the particular job.[72] Again, the MSAs, or mission support areas, fall under the control of different entities, which set their own rules for operation.

The following is one example of an MSA. The Mariner missions of the 1960s and 1970s had 640 square meters in SFOF for operations.[73] The MSA, Room 111 on the first floor of the SFOF, included a large primary support area and four smaller rooms. In the main area, at least ten workstations were aligned in five rows. Printers, storage cabinets, and other necessities skirted the perimeter of the walls. Room 111A, the conference room, contained a long, rectangular table in the center of the room and a chalkboard, bulletin board, and map of Mars on the walls. Room 111F, the Mission Director's Room, had three workstations, a bookcase, a bulletin board, and a chalkboard, as well as the NASA phone. A large, circular table sat in Room 111G, the Mission Control Room. This room also had numerous storage cabinets and a printer. Finally, Room 111H, the Observation Room, included a rectangular conference table and display board. Mariner also utilized a Spacecraft Performance Analysis Area, located next to Room 111. This area housed about thirty workstations, displays, storage cabinets, and a small conference room. A Principal Investigator's Area lay on the other side of the Mission Support Area. This room held around twenty-five workstations, storage, displays, and a conference table large enough for ten chairs. Finally, the Flight Path Analysis Area was located adjacent to the Spacecraft Performance Analysis Area. This room included twenty-five consoles, plotters, storage cabinets, and numerous televisions.[74]

One final difference between JPL and other centers must be mentioned. Controllers working on Mars missions observe the Martian day, or sol, which lasts forty minutes longer than a day on earth. While this proves invaluable to the mission, it can cause short-term problems for the individual controllers. Many can become disoriented and so focused on their new time cycle they have no concept of earth time. One controller even mentioned that after nearly falling asleep at the wheel while driving home after a shift, she brought a sleeping bag and pillow to her office to prevent another such occurrence.[75]

JPL's Operations Room differs greatly from JSC's MOCRs and FCRs. Rather than the main control room handling the majority of the work, for JPL's missions, the individual MSAs control the missions, with few exceptions. A typical mission works out of the main Operations Room only during launch and other critical elements. The Operations Room generally remains unused, aside from the DSN controllers. JPL has also optimized the room to control multiple missions simultaneously, while JSC's control rooms generally only monitor a single spacecraft.

CONTROL ROOM WORK

Much like those at JSC, controllers at JPL must endure intense training for each mission. Simulations are equally necessary for the unmanned missions controlled at JPL. Like the JSC controllers, those at JPL spend ten or more times as many hours training and simulating missions than at the console for the actual missions.

Aside from the usual computer simulations, JPL engineers preparing the various Mars rovers have undertaken extensive simulations with model rovers. They can only use models, since the engineers can build only one real rover and it must be maintained in a clean environment. The models, however, are as exact copies as possible, and the engineers seek to duplicate the conditions of the environment to which the rovers will be sent as accurately as possible. Scientists and engineers have sought out various natural land formations thought to be similar to the surface of Mars. To demonstrate the possibilities of a rover on Mars, engineers drove a rover in the Arroyo Seco near Pasadena.[76] A model *Sojourner* ran through numerous tests in the Channeled Scablands in eastern Washington state. NASA scientists have deemed Iceland a Mars analogue since the Viking program of 1976. Numerous scientists preparing for subsequent Mars programs have visited the most desolate areas of the country in an attempt to anticipate what they will encounter through the rovers on the distant planet. At least one scientist also journeyed to the remote Navassa Island between Jamaica and Haiti.[77] In fact, these treks call to mind the geological expeditions of the Apollo astronauts before their visits to the lunar surface.

Most rover models have also gone through extensive testing in man-made environments. For *Sojourner*, for instance, JPL prepared a room with sand and various rocks, which, appropriately enough, they called the sandbox. With curtains drawn to prevent outsiders from looking in, personnel rearrange the rocks to present a different test surface. Controllers then must use the onboard cameras to take images of the environment and move the rovers around just as they would on the Martian surface, complete with a time delay.[78] Such real-life simulations can prove invaluable to the engineers and controllers as they prepare for upcoming missions.

For the *Sojourner* project, JPL first built mock rovers of various sizes, from eight inches to the size of a truck, to prove the feasibility of the mission. In this case, the primary test vehicle was the System Integration Model (SIM), better known as *Marie Curie*. After they built the actual rover, it underwent months of tests. Full tests must include simulated sun and stars, light, and temperature. In the month before launch, after the spacecraft had been placed on the launch vehicle, JPL technicians tested the DSN to ensure that communication links worked. Controllers, meanwhile, simulated practice countdowns. Following the launch, the spacecraft flew in transit for approximately six months, during which controllers and engineers completed their final simulations. They utilized both computer simulations and model rovers in the sandbox. The operations team even conducted more field tests of *Marie Curie* in the Mojave Desert. By this time controllers had already transferred to the slightly different Mars time in preparation for on-surface activities. When the spacecraft reached its destination, the engineers and controllers were ready for virtually any potential problem.[79]

All control rooms must put the training and simulations to use at some point, and JPL is no exception. Unfortunately, the most famous failures in robotic spaceflight history, like the 1990s Mars missions, have been so catastrophic that fixes were impossible. Yet controllers have been able to come up with alternatives in some cases. One example of controllers working around problems came with the launch of the two Voyager spacecraft. Shortly after the launch of *Voyager 2*, the science boom failed to fully deploy. Fortunately, the spacecraft and its experiments continued to work properly. More troubling, the orientation of the spacecraft erratically changed without warning. This not only caused problems for alignment but also expended propellant at a higher rate than expected. Controllers determined, however, that neither of these issues would cause a mission abort.

Meanwhile, engineers had time to learn from those mistakes and fix some of the problems with *Voyager 1* before its launch. Interestingly, the designers had given the spacecraft so much autonomy that the controllers could do little more than watch and hope as *Voyager 2* tried to correct itself. JPL personnel also ran into some issues as Voyager flew. Some controllers began to work on the next proposed project: Galileo. Others were so consumed with correcting problems with the Voyager spacecraft that they fell behind in their planning for the planetary encounters. Controllers had to update the onboard systems continually as the spacecraft flew farther away from the earth and the sun. After some reorganization and an increased budget, Voyager eventually outperformed its original goals.[80]

Missions to Mars have overcome some near disasters as well. For example, after *Mars Pathfinder* arrived on the Martian surface, the controllers recognized a potential problem. The *Pathfinder* lander served as a cocoon surrounded by airbags. The airbags softened the landing, but still it landed with such force that it bounced more than a dozen times before coming to a complete stop, as designed. Once the lander stopped, the airbags deflated. Then the capsule opened like a flower, and the rover could drive down using a petal as a ramp. Unfortunately, the airbags did not properly deflate and instead surrounded the petal *Sojourner* would drive down. During emergency simulations, the rover caught on the airbags, which then covered the solar panels on the rover and caused it to die. Using information gained from those tests, controllers sent commands to raise the petal, retract the airbags, and then redeploy the petals. This maneuver allowed the mission to proceed as planned.[81]

Controllers must also adapt to unexpected problems. In 2009, five years after landing on Mars, the rover *Spirit* became stuck in the Martian soil. The team attempted to free the rover for almost eight months, testing various procedures in the JPL sandbox. With the Martian winter rapidly approaching, JPL decided to save the remaining solar power and stop extrication efforts. Instead, the team adapted and used *Spirit* as a stationary platform, conducting experiments otherwise impossible with a mobile rover. For instance, it was able to test the planet's wobble, perform a concentrated study on the nearby soil, track the movement of particles by wind, and observe the atmosphere.[82] The rover continued its mission for two months until JPL lost contact with it on 22 March 2010. Controllers persisted in sending signals to no effect for over a year until JPL officially ended its mission on 25 May 2011.[83] In the end, the rover's operations lasted more than six years, well past the planned three months. *Opportunity* continues its mission on Mars.

*   *   *

As with Mission Control at JSC, JPL's control center played an integral role in NASA's efforts to win the Cold War. Incorporated as a laboratory with an impeccable history of work with engines and rockets, JPL immediately lent its expertise to the nascent space program by building the first American satellite. In this way it set its own precedent for space exploration: robotic spacecraft on the vanguard, paving the way for human spaceflight. The Apollo program could not have reached its goal of landing on the moon before the end of the 1960s without the pioneering missions of the Ranger and Surveyor programs. JPL not only built those spacecraft, but also, crucially, served as Mission Control for the first American missions to the moon.

JPL and its control room differ from JSC. As a laboratory already in use and later added to NASA, the space agency had no control over its location. Its history also means that it has a close, sometimes contentious, relationship with Caltech and the Army. The relationship with the Army continued throughout the Cold War, and while it became less prominent in the 1990s, a military aspect of the JPL mission remains.

Missions are controlled in a strikingly different manner as well. The Main Operations Room, aside from DSN Control, is not manned continuously during missions. Instead, the majority of work is completed in Mission Support Areas. There is no true equivalent to the flight director of JSC. Instead, managers consult with the controllers and other experts, building a consensus before decisions are made.

There remain important similarities between the two, however. The Mission Control buildings have a centralized location within the space centers. The Main Control Rooms have almost always utilized consoles. They also have viewing screens at the front of the room, projecting images and including a mission-duration clock. The rooms also include viewing areas for the public to see the controllers at work. NASA has continuously updated the control centers as technologies and missions evolve to ensure survival, particularly as budgets declined. With NASA's concepts of mission control in mind, it is important to understand how another space agency has constructed its own control center.

3

EUROPEAN SPACE
OPERATIONS CENTRE

Darmstadt, Germany, is a study in contrast with the homes of the American control centers. While Johnson Space Center is situated thirty minutes from the heart of Houston, it is surrounded by bustling suburbs. The Jet Propulsion Laboratory, located a similar distance from Los Angeles, is ringed by residential areas and lies just off a major highway. Both can be reached easily by the public, are in or near two of the four most populous cities in the country, and often are the center of attention for spaceflight news.

Darmstadt, on the other hand, is not even among the fifty most populous cities in Germany. There are only a handful of restaurants in the town, and few cater to tourists. Hotels are sparse. The area around the control center feels starkly industrial, a feeling heightened by its proximity to the train station. The train station, however, provides one hint of modern convenience, since it is the first stop on most express trains on the southern route out of Frankfurt, Germany. Since Frankfurt is the fifth-largest city in Germany and hosts an airport that is a major hub for central Europe, it could act as a funnel for visitors to Darmstadt, but the town remains a stop primarily for those with business either with the European Space Operations Centre (ESOC) or the Technische Universität Darmstadt. Unlike Los Angeles and Houston, Darmstadt rarely sees vacationers.

In many ways, Darmstadt is the perfect location for the ESOC. Whereas JSC and JPL, throughout their histories, have fought to remain visible to the public, for much of its existence ESOC has intentionally maintained a low profile. Leo Hennessy, head of personnel during the mid-1990s, likened ESOC to the defenders of a castle. They wanted to keep their heads down and survive. If they stuck their necks over the wall too often, someone might notice and knock them down. Only recently has the European control center felt secure enough in its position to have a stronger presence in the public eye.[1]

While the location may differ significantly from the other two mission control centers, JSC and JPL, the European Space Operations Centre does possess remarkable similarities to the other two centers. Space agencies constructed each center for a unique mission with a unique history. Each also was built for specific reasons during the Cold War. Before this chapter presents a more detailed look at the control room, it must provide the historical context for the center's existence.

EUROPEAN SPACE OPERATIONS CENTRE

Unlike NASA, the European Space Agency (ESA) represents not one but many European national space efforts. It, therefore, must recognize the needs and the differences among its constituent countries when making decisions about operations. It has also always operated with a budget a fraction of the size of NASA's and must limit the size and number of missions or actively search for partners for larger projects. One other major difference arises from the lack of a substantive human spaceflight program—until recently, that is. It has never launched a spacecraft able to accommodate humans, and thus relies on other space programs to carry its astronauts into space. Since ESA remains primarily a robotic spaceflight program, with a substantial deep-space history, its operations resemble JPL's more closely than JSC's.

In order to understand the makeup of ESA, one must first understand its roots in the immediate post–World War II era. Across Europe the war had caused devastation and destruction on the land and infrastructure. Following the war, European science and technology, especially in Germany, suffered a significant "brain drain." Rocketry pioneer Wernher von Braun, and about 120 other engineers, moved to the United States to aid its early rocketry efforts. The Soviet Union relocated another 200 engineers and scientists, most notably Helmut Grottrup.[2]

Western European nations concentrated on newly emerging technologies as a way to rebuild their economies quickly and regain their standing on the international stage.[3] The European Organization for Nuclear Research (CERN) was just one of the many agencies to be founded during this time. After some years of discussion, twelve nations agreed, in 1954, to found the particle physics laboratory in Geneva.[4] CERN remains an important international organization that has continually grown over time.[5]

In the late 1950s, a number of European scientists, many of whom were closely tied to CERN, began to promote a European space organization. Edoardo Amaldi, an Italian physicist, and Pierre Victor Auger, a French atomic physicist, were two of the main catalysts for this movement. By 1961, they had successfully convinced a number of Western European governments to cooperate in creating the European Preparatory Commission for Space Research (COPERS). Out of COPERS came a proposal for a European Space Research Organisation (ESRO). By its third meeting, in October 1961, COPERS had created a document referred to as the Blue Book, which outlined ESRO's organization, its purpose, and the technology it needed to function. Many of these early ideas became reality.[6]

On 20 March 1964, Belgium, Denmark, France, Italy, the Netherlands, Spain, Sweden, Switzerland, the United Kingdom, and West Germany officially established ESRO.[7] Europeans realized early on that individual nations did not have the financial resources to compete with the United States and the Soviet Union. The European nations cited previously created ESRO to counter, but not compete against, the space agencies of the United States and the Soviet Union. By working together in ESRO they could give scientists and engineers a reason to stay in their home countries and not emigrate to one of the major powers, thus avoiding a continued "brain drain." In addition, the European authorities recognized that involvement in space might lead to advancements in technology that could boost the economic and industrial development of their nations.[8]

As European officials met to discuss the creation of ESRO, a significant difference of opinion arose about launchers for potential satellites. Most agreed that the Europeans needed to build their own rockets to avoid dependence on other space agencies. Because it was up to Britain and France, the more highly industrialized nations, to take charge of launcher production, they wanted a separate agency to manage the production and utilization of launchers. Others, particularly Belgium, and to a lesser extent the Netherlands, Switzerland, and Italy, argued for one agency to oversee all space operations, the political persuasion of Britain and France won out. This led to the creation of the European Launch Development Organisation (ELDO) as a separate entity.[9]

The separation of space efforts did not last. On 31 May 1975, as part of a new reorganization, ESRO and ELDO combined to create the European Space Agency (ESA). (For the purposes of simplification, this work will use ESA to refer to the European space agencies, regardless of date).[10] The Convention for the Establishment of a European Space Agency, which oversaw the creation of ESA in 1975, stated that ESA's purpose was "to provide for and to promote, for exclusively peaceful purposes, cooperation among European States in space research and technology and their space applications."[11] This statement affirmed a commitment to peaceful, nonmilitary missions. Later in the document, Act XIV discussed cooperation with outside parties. A unanimous vote by all member states is required to approve any work with a foreign company or space agency.[12] All member states recognized cooperation as an integral aspect of the agency.

While the member states must be unanimous in their approval of foreign cooperation, they do not always agree on which programs to pursue. As a result, Act IV of the conference described the funding of activities. ESA labeled all programs either "mandatory" or "optional." Funding for mandatory programs came from all member nations in proportion to their gross national products. Each nation had a single vote on all critical matters, and each issue required a unanimous vote. Mandatory programs included technology research, education, facilities, solar system science, astronomy, and fundamental physics. Member states chose their level of involvement in optional programs, which included earth observation, telecommunications, satellite navigation, space transportation, the International Space Station and human spaceflight, robotic exploration, and microgravity research.[13]

ESA's industrial policy, outlined in Act VII, stipulated that programs had to be cost-effective, yet competitive, with equitable participation by member nations, and a preference to utilize the industry of member nations. Finally, ESA had to pursue free competitive bidding on projects, unless that interfered with the other requirements.[14] Some of those rules caused problems for officials deciding on the contracts. It can be quite difficult to sync equitable distribution of funds with free competitive bidding and cost-effectiveness, but ESA did its best to adhere to the guidelines.

Interestingly, Act IX stated that any member could use ESA facilities for its own non-ESA programs, as long as their use did not interfere with regular ESA programs.[15] This allowed for another source of income for ESA. More fundamentally, this idea of openness had been important for ESOC in particular as it emerged on the international stage as a major control center, since it had allowed ESOC to participate in rescue activities for non-ESA spacecraft. This openness could likewise encourage the spread of technological ideas considered essential to the recovery of Europe. More recently, ESOC has chosen to rent its facilities to any user who might need them. The administration cites both its proven expertise and the flexibility of its facilities when promoting the center to international customers.[16]

The ESA Council, which included two representatives from each member state, ran the general business of the space agency. The council appointed a director-general to serve as the highest-ranking official for ESA, similar to NASA's administrator.[17] In essence, if the ESA Council were the board of directors, the director-general was the CEO of ESA.

The European Space Agency established a number of centers across the member states for various aspects of the organization's activities. ESA accepted several bids for the centers. France, Switzerland, and the Netherlands, for instance, expressed interest in the headquarters. The

European Space Research and Technology Centre (ESTEC) proved to be the most sought-after center, with bids from West Germany, France, the United Kingdom, Switzerland, Belgium, Italy, and the Netherlands. On the other hand, the European Space Data Acquisition Centre (ESDAC) initially received only one bid, from West Germany, though the United Kingdom and Switzerland later expressed interest. ESA officials weighed many variables while making their decisions on locations, including cost and efficiency, though perhaps the most important criterion was political considerations.[18]

On 14 June 1962, the Conference of Plenipotentiaries resolved the locations for the various centers.[19] Built to handle spacecraft development, the European Space Research and Technology Centre (ESTEC), in Noordwijk, the Netherlands, became the largest and perhaps most important center. In addition, it housed the control center for the satellite tracking and telemetry network, ESTRACK, which originally had four ground stations, in Redu, Belgium; Fairbanks, Alaska; Spitzbergen, Norway; and the Falkland Islands.[20] ESTEC was responsible for the second phase of programs, that is, the transmission of data from spacecraft to the ground and orders from the ground to spacecraft.[21] ESA originally selected Delft, in the Netherlands, for its location, but this soon changed to nearby Noordwijk. Kiruna, Sweden, hosted ESRANGE, the sounding rocket launch area. Darmstadt, West Germany, became the site of the European Space Data Acquisition Centre (ESDAC), with the large-capacity computers used for calculations and for the study of spacecraft data.[22] The original tasks of ESDAC included processing and analyzing data, performing orbit computations, and conducting scientific work on data from experiments. The facilities did not allow for any real-time computation.[23] During the debate regarding the location of the centers, eight members voted to situate ESDAC in Darmstadt, while four others voted for Geneva, Switzerland.[24] Paris became headquarters for ESA. Finally, the European Space Research Institute (ESRIN) in Frascati, Italy, eventually became the lead center for earth observation.

ESDAC originated as a few offices in a building run by a computer company, Das Deutsche Rechenzentrum (German Computer Centre) on Rheinstrasse, in the research and technology district of Darmstadt. In early 1964 it moved to a new office nearby, in the Deutsche Buchgemeinschaft (German Book Society) on 16 Havelstrasse.[25] From its start, ESDAC claimed to have "one of the largest and most modern computer installations" anywhere.[26] Originally, ESDAC comprised the data processing division and the data analysis division.[27] Mission analysis included the mathematical examination of satellite orbits, which informs the design of the satellite in its early developmental stages.[28] ESDAC had thirty-one employees in 1965 and forty-nine by the end of 1966, and the center's eight-year plan forecast a staff increase to eighty in the new facilities.[29] The staff complement had risen to 295 in 1969 and to 311 in 1970.[30] In 2010, of the 2,072 total ESA staff, 244 worked in Darmstadt.[31]

By 1966, ESA executives realized that the organization was working inefficiently and needed an overhaul. As a result, they created a group, led by Jan H. Bannier, director of the Netherlands Organisation for the Advancement of Pure Research (ZWO) and former chairman of the European Organization for Nuclear Research (CERN), in the hopes of solving these problems. Perhaps the most important and longest-lasting result of the Bannier report was the recommendation that ESA situate the control center closer to the data locus, ESDAC. This created a central location for spacecraft data and reduced redundancies in equipment and personnel. Thus, the operations control center moved from Noordwijk to Darmstadt, and ESA renamed ESDAC the European Space Operations Centre, or ESOC. (For clarification the acronym ESOC will be used for the remainder of this work.)

ESOC serves as a vital link between spacecraft and the end users.[32] It focuses on the calculating and processing of data. ESOC continually strives to remain at the forefront in computer technology in order to maintain its presence in the space program. As an example of the central role of computing technology within ESA, the ESOC director originally held responsibility for agency-wide computer utilization.[33] ESOC's history is one of continual technological change amid remarkable constancy in focus. Preserving a high standard of excellence in technology was especially important during the Cold War as

ESA balanced competition and cooperation with the two superpower space agencies. This has allowed ESOC to continually update its hardware and other technology while maintaining its success in matters of control.

Howard Nye, a spacecraft operations manager and flight director, described the mandate of ESOC as being to "conduct mission operations for ESA satellites and to establish, operate and maintain the necessary ground segment infrastructure." He went on to explain mission operations as a "process involving operations planning, satellite monitoring and control, in-orbit navigation, and data processing and distribution."[34] These two definitions given by Nye summarize the work of ESOC and, especially, the Operations Control Centre (OCC), for its more than forty years of existence. In that time, it has had three primary duties: planning, operating and maintaining ground segments, and data processing.

With its impending expansion as ESOC in 1966, ESA constructed new buildings and facilities in a field west of town and just west of the railway station. The federal government of West Germany gave ESA the land for a ninety-nine-year lease with a one-time fixed price of 1 deutsche mark per year, a prior right to renew the lease, and a stipulation that the lease would end early if ESA were dissolved.[35] The contract further stipulated that the land could be used only for ESA purposes, and that if the contract did expire and the land revert, the state would compensate ESA for any permanent buildings. Finally, it stated that ESA must relinquish any natural resources or historical pieces found on the property to the local government with no compensation.[36] The terms of the ESOC contract echo many of those in JSC's contract with Rice University. In order to use that land, ESA had to negotiate further with national and local authorities to clear trees as well as build an access road.[37] Almost immediately, ESA realized that ESOC needed more space. By October 1969, the local government had agreed to lease an additional twenty thousand square meters for ESOC.[38]

At its new location ESA upgraded its main computer to an IBM 360/50, installed and placed online in September 1966.[39] Although ESA originally scheduled completion of the new facilities in mid-January

1967, ESOC did not become fully operational until 17 May 1968.[40] Interestingly, the local newspaper, *Darmstadter Echo*, referred to ESOC as the "Houston of Europe" on the first day after commissioning.[41]

ESOC originally consisted of two main buildings. The headquarters, or administration building, included three floors with a partial basement for a total of 1,700 square meters. This building housed offices, a library, a conference room, and an area for visiting scientists. The computer building was 1,200 square meters on the ground and basement levels for various services focusing on the computers.[42] ESOC controlled the first missions from converted administration office space because it did not have any dedicated rooms for the control center at that time. The original control room consisted of seventy-two square meters of office space in a corner of the second floor of the administration building.[43] Those displaced by the control center worked out of a temporary wood building in the parking lot. By the end of 1967, ESA transferred the telemetry data-processing line, interface equipment between that line and the computer, the teleprinter communication links with ESTRACK ground stations, and the equipment for the network operations room to the new ESOC facilities.[44]

ESOC transferred all major control equipment to the new facilities in time for the Aurorae satellite mission in May 1968. During the Aurorae mission, ESOC's responsibilities included coordinating the work of the stations, as well as initiating and processing data. Its primary concern was maintaining the flow of data.[45] From the beginning, it operated more similarly to JPL than to JSC.

In 1968, ESOC established a protocol for the use of its computers by outsiders. It did not charge experimenters working in cooperation with an ESA project. It charged all others a nominal fee for the use of the equipment, staff pay, the full cost for consumables, and a service charge. ESOC added that outside users could use their own operators if they were fully qualified, thereby potentially reducing their cost.[46] Outside users could include members of various technological companies in Europe or members of other space agencies from around the globe. Even when it came to hardware use, ESA preferred cooperation over competition, particularly during the Cold War.

By the end of 1969, ESA realized that the current setup was difficult to use and insufficient for the agency's needs. That ESOC's area was undersized and not originally built for the purposes of a control center made a difference as well. ESOC further reasoned that a new control center made ESOC more marketable for outside projects.[47] Construction on a new Operations Control Centre (OCC) began in February 1970. After its completion in March 1971, the administration moved from their temporary quarters permanently into their offices. ESOC was now fully operational in accordance with the original plans.[48]

In 1971 the ESA Council released a report detailing a number of recommendations on the structure of ESOC. They began by recommending simplification of the administration. The council argued for a stronger operations management group. It further recommended that a tracking and data systems manager be named for each spacecraft as part of the project team directly responsible to the project manager. Finally, the council stated that ESOC should be responsible for the equipment at all of the ground stations in the network, the Control Centre, and the launch ranges. The ESA administration quickly approved this report and made the appropriate changes.[49]

During the mid-1970s, ESOC also planned for the construction of a new three-story building with 3,250 square meters of total floor area, largely for the needs of the new Meteosat program. It included areas for the Meteosat computer system, control room, work areas for meteorologists, as well as a room for the new European computer system.[50] In 1980, ESOC installed electronic locks across its campus to control access to its buildings. Employees insert a card to unlock doors to buildings and critical areas within the buildings. They were able to store information concerning who was unlocking what doors, in case they needed that information in the future. The system did not always work, especially immediately after installation.[51]

More facilities came in the late 1980s and early 1990s. In 1989 ESOC completed the first part of a building designed for extra offices and workshops, though not those directly connected with the OCC. Two years later ESOC added a 355-spot, multistory parking garage as well as a new energy center for emergency power. In 1995 contractors completed construction of Building E, which originally housed dedicated control rooms (DCRs) for the ERS/Envisat, Cluster, and Huygens programs, as well as areas for the Cluster principal investigator (PI) and electrical ground support equipment (EGSE).[52] In 1993, ESOC built a new flight dynamics room in the OCC to take advantage of new hardware upgrades.[53]

The International Organization for Standardization (ISO) is a non-governmental organization, formed in 1946, that seeks to establish standards for the workplace. The 9001 certificate recognizes quality assurance.[54] In recognition of its impressive track record, ESOC received ISO 9001 certification on 30 November 1999. It was the first ESA center to receive such an honor.[55] ESOC rightly takes pride in this award, and other ESA centers have attempted to follow suit.

As an international organization, ESA drew upon its many European constituents, and even from outside Europe, for its workforce. It had, therefore, a diverse working environment. Each center was equally diverse, and ESOC was certainly no exception. In this way it contrasted strikingly with the control centers of NASA, which employed Americans almost exclusively. Not surprisingly, there are differences in how controllers worked in such a multinational environment, compared to those in the United States.

Most Europeans consider working in an international organization normal. ESA was not unusual in that regard. Employees from different countries worked so well together that they tended to identify people by their departments within ESA than by their nationalities.[56] By acknowledging that everyone must work together in order to accomplish their jobs, individuals could easily forget any differences stemming from national pride, even during the World Cup soccer competitions. In fact, the majority of ESA enjoyed a high level of teamwork.[57] Former ESOC director Felix Garcia-Castañer attributed his center's success to the dedication and enthusiasm of the employees.[58]

When compared with the locations of other ESA centers, such as Paris and Rome, Darmstadt was less well known internationally. Even with Frankfurt relatively nearby, ESOC seemed at times distant and apart from the rest of the world. Despite that, during a ceremony commemorating the laying of the foundation stone for the new ESDAC building on 12 November 1965, Dr. Gerhard Bengeser, the assistant director for administration at ESDAC, stated that he felt that "most of the staff members feel themselves at home now in this charming city which is a heaven [*sic*] of Science and the Arts."[59]

If they were not German, new employees adjusted not only to a new place of employment but a new country and culture as well. ESA employees have expressed experiencing at least minor culture shock when moving to a new area.[60] The transition could be especially taxing for families, particularly if a different language was spoken in the new place. While the employee spent much of his or her time working and speaking in English or another common language, the family had to adjust to what could be a major language barrier. Often, parents experienced difficulties helping their children with schoolwork or speaking to their children's teachers, for example.[61]

ESOC as a whole grew from the original two thousand square meters of building space to almost thirty thousand square meters in the 1990s.[62] More than just a workplace, it served as a hub for employees from the many different European countries to come together and perhaps even bring some cultural elements of their home countries with them. ESOC had a plethora of clubs and organizations for its employees and families to give them an outlet the local community might not have been able to provide. Their diversity also showed the human side of ESOC. At any one time, there had been up to thirty-eight different clubs for various interests, including singing, golf, theater, and soccer. There were also organizations for various cultural groups represented in ESOC's staff.[63] The Ladies' Club was one of the original clubs, organized to help wives of ESOC employees in their transitions. Eventually the club was subsumed by the Social Committee. The ESOC Canteen, opened in 1970, hosted many of the social events for different clubs and organizations.[64] Before it closed in 1987, the bar also served as a meeting place, or a watering hole, for employees.[65]

Not all women associated with ESOC are relegated to the Social Committee. In fact, women have long played a major role in ESOC operations. The author of the premier ESOC memoir, Madeleine Schäfer, began employment at the center only a few months after its inception. The 1969 ESRO General Report included a photograph of a female operator in the Darmstadt facilities.[66] This suggests that ESOC included female controllers well before either of NASA's control centers did. As a member of an international organization, ESOC had to integrate from the start, regarding both gender and nationality.

ESA's 2010 budget was about 3.7 billion Euros, of which ESA returned nearly 90 percent to European industry in various ways, including research and development. ESA ensured that, to the best of its abilities, there was fair distribution of research and development funding to each member state. The ESA Council established a new budget every three years. This multiyear approach allowed for better planning for future missions. ESOC received roughly 380 million Euros of the 2010 budget.[67]

Contractors were another way that ESA returned the investments to the countries. ESA often took into account the national origin of contractors before making decisions as to whom they should hire, although the ESRO Convention specifically stated that scientific, technical, and economic factors should take precedence over geography.[68] Many contracts were awarded to the largest contributor of a particular project. For instance, Germany provided funds and labor adding up to 52.6 percent of Spacelab, France 62.8 percent of Ariane, and Great Britain 56 percent of MAROTS.[69] ESA did its best to adhere to the ideas of *juste retour*, an industrial return coefficient that measured the amount of return compared to the amount of input.[70] This method meant that ESA sometimes could not accept the best proposal, depending on the company's country.[71] Sometimes, though, the makeup of the contractor made it difficult to define the home country or to determine whether the business was from a member state or not.[72] Because contractors were hired only for a specific task, ESA could spread around the money to different countries. More so than NASA, ESA must constantly keep politics in mind while also attempting to provide the best services possible.

Limited contracts also allowed ESA to keep its permanent employment to a minimum. The individual contractors sometimes found ways to work around the limited employment. Some switched firms in order to remain at ESOC for longer periods of time, which could cause confusion for employees and for records that kept track of each individual's work status.[73] Sometimes contractors also joined ESA as staff members.[74] After all the calculations, ESA maintained that *juste retour* had successfully encouraged the member nations to contribute to the European space program.[75]

OPERATIONS CONTROL CENTRE

When ESA constructed the European Space Operations Centre in Darmstadt in 1966, it controlled missions in a series of makeshift offices in the administration building. Realizing the inadequacy of this arrangement, ESOC built the Operations Control Centre (OCC) as the new focal point of the organization. It began operations with the *Thor Delta (TD-1)* satellite in 1972.

The OCC originally covered a total of nine hundred square meters of floor space, including main and auxiliary control rooms, an experimenter evaluation room, and an orbit operations room.[76] It also contained a training room for controllers, offices for project representatives, equipment checkers, network and spacecraft controllers and officers, an ESTRACK communications lab, and building systems.[77] The OCC included a 140-square-meter network operations room with two adjacent project operations rooms of comparable size. This setup mirrored those of NASA and the original control center at Noordwijk and maximized efficiency during any extended phases of missions. A two-hundred-square-meter viewing room attached to the operations room doubled as a lecture room or a training area when the room was not being used for the Launch and Early Operations Phase (LEOP).[78] Like JSC and JPL, the OCC also housed diesel motors for emergency power.[79]

The OCC also included a fifty-five-square-meter communications room, eighty-square-meter controller offices near the operations room, a one-hundred-square-meter area for visiting project staff and scientists, a similarly sized room for the operational project staff offices of projects for other organizations, a fifty-five-square-meter auxiliary operations room with additional equipment, fifteen-square-meter offices for the head of the network and the head of communications, twenty-five square meters for a kitchen and restrooms, and a sixty-five-square-meter display projection room for the rear projection system.[80] Hardware included two IBM-1802s with alphanumeric displays and hard-copy devices, a computer interface system, analogue tape recorders, a teleprinter, and closed-circuit television (CCTV).[81] ESA constructed the operations building to control up to four spacecraft simultaneously.[82] After its completion in March 1971, the administration moved permanently back into their offices from their temporary wooden building, and ESOC was fully operational in accordance with the original plans.[83]

An IBM type 2260 AND system, installed in early 1970, allowed the controllers to view telemetry displays for the first time, rather than relying on the sporadic orbital determinations computed previously.[84] After the Operations Control Centre's completion in March 1971, the contractor handed over the principal equipment configuration for control to ESA in July. Over that summer, ESOC conducted extensive staff buildup and training, along with a reconstruction and reorganization of the computer center, in preparation for the launch of *HEOS-A2*. While it was originally scheduled for launch in December 1971, it was delayed, due to various concerns, until January 1972.[85]

In preparation for new computing requirements in the mid-1970s, ESOC conducted a general overhaul of its computing facilities between 1972 and 1974. In August 1972, they replaced the IBM 360/65 with an IBM 370/155, which included more central memory (1.5 megabytes) and more disk storage (1.2 megabytes) at a lower cost. This new mainframe also allowed for computer "time sharing" for the first time in ESA's history.[86] During this time ESA officials also planned to install two ICL 4/72 computers, one each at ESOC and ESTEC, as part of an offline system. They shared 4,800 bits per second data link. This allowed ESOC to have a one-megabyte core storage memory and 580-megabyte disk storage capacity. They originally planned for these computers to become operational by January 1975, and the new IBM was maintained until 1977.[87] By the end of 1973, ESOC had also installed an IBM 1800, so that while the 370/155 processed data transmitted from spacecraft, the 1800 operated in real-time for data recovery from spacecraft.[88]

In 1973, ESA also agreed to allow the Netherlands Space Agency (NLR) to use the OCC and the Redu ground station to control its ANS mission, which flew from August 1974 to March 1976. ANS proved to be the first ESOC experience with onboard computers, which was not

duplicated again until Exosat was built in 1983. In order to accommodate this new technology, ESA installed a STAMAC communication
interface system to allow for real-time keyboard telecommand.[89]

GEOS-1, in 1975, included no onboard computers, so 90 percent of
the data was downloaded to the ground in real time. The process was
fully automated, both onboard the satellite and on the ground, for the
maximum scientific output. With a new fully automated ground system, ESA had a capability no other space agency could duplicate. The
ground and satellite shared about 25,000 commands per day with a
maximum delay of only ten seconds. They had only 3,500 preplanned
commands, because they had to continuously adapt to changing atmo-
spheric conditions. GEOS-2 flew in 1977 in a mini constellation with
GEOS-1. This demonstrated capabilities that proved influential for the
later Cluster mission.[90]

As part of its commitment to aid the development of European technological industry during the Cold War, ESA tried at various times,
beginning in 1969, to replace their IBM systems with computers from
Europe. After one more unsuccessful attempt in 1970, they placed a
new request in 1971. They received three main bidders: IBM, CII + Siemens, and "Group One," a consortium including the European companies International Computers Limited (ICL), AEG-Telefunken, SAAB,
CSI, and Rousing. After long negotiations, the administration agreed
on a mixed layout, including real-time computers by CII + Siemens
and batch computers by ICL. They also signed a separate contract for
software with Rousing. Under this contract, the real-time equipment
was installed by the end of 1974, though the IBM systems remained
online until 1977 due to the needs of the COS-B satellite.[91] The majority
of these systems did achieve acceptance on time.

Installation of the ICL 4/72 computer for batch processing began in
September 1974, and the machines completed handover tests in December. CII installed one 10070 computer for real-time work in March
with acceptance in June, while a second was installed in August for
redundancy. The seven Siemens 330 computers, which were planned
to aid real-time work, experienced problems, however, and could not
become operational until January 1975. This new European system of
computers led to the Multi-Satellite Support System (MSSS), which

was installed in 1975 in anticipation of the OTS, MAROTS, and Aerosat programs.[92] The MSSS, fully completed in 1976, allowed for the tracking of up to six satellites simultaneously.[93] This new facility supported LEOP for any and all missions while also providing an area for the routine phases of missions that did not need a specific dedicated control room, such as OTS, MARECS-A, and GEOS. The MSSS included eight Siemens 330 primary processors and two SEL 32/7780 computers for their system.[94]

In 1975, ESOC also modified the Operations Control Centre's main room for newly launched spacecraft only. The administration decided to build dedicated control rooms (DCRs) for GEOS and Meteosat, as well as for future long-duration missions.[95] Due to changing needs and updated technology, by 1977 the Meteosat Ground Computer System (MGCS) included two ICL 2980s, six Siemens 330s, three Rousing CR80 Array Processors, and two Data General Nova 830s—an impressive set of components.[96]

New computer hardware installed in 1978 was supposed to continue functioning until 1985, but ESOC recognized that major changes in computing required upgrades earlier. The price of computer facilities dropped rapidly, though the price for manpower rose steadily.[97] During the 1980s ESOC continued to upgrade its computer systems to keep up with technological advances. For instance, in 1980 the two CII 10070s were replaced with two SEL 32/77s for use during simulations.[98] They also installed a DPS 66/05 system for general purpose, shared computing.[99] The following year the MGCS received an upgrade from its ICL 2980s with two Siemens 7865 computers backed by an IBM 370-148 and a CII-HB DPS/05 offline system, due to its better price and performance.[100] With the installation of the new MGCS, ESOC, for the first time, used representative spacecraft models as test data sources to validate the ground support systems. This quickly became standard practice when installing new systems or preparing for new missions.[101]

By 1983, the Siemens 330s in the MSSS were obsolete and accrued high maintenance costs, so they were replaced by two Gould/SEL-32/6750 mainframes, one of which was used exclusively for backup.[102] The new mainframes of the 1980s included one major technological change: they did not use punch cards. Users instead typed programs directly into terminals. While the mainframes may have been more user-friendly, ESOC did experience some problems with disk space. In fact, a logbook for failures and crashes associated with the new mainframes was quickly abandoned because there were too many documented for the engineers to keep up with them.[103]

Between 1983 and 1985, ESOC modernized the OCC display and control facilities. The previous consoles, which had been installed in 1971 and modified in 1976, required replacement for the new nineteen-inch color monitors, three for each console, and standard keyboards with function keys.[104] In 1986 and 1987, ESOC added two stories to the OCC to accommodate the Giotto project and staff.[105] More major computer upgrades came in 1990 with the installation of three new mainframes. The Comparex 8/90 handled mission analysis, payload data processing, and other similar programs. A Comparex 7890F supported Meteosat Data Processing and Flight Dynamics. An IBM 4381/R14 was used for general-purpose computing as well as office automation.[106]

In the late 1980s, ESOC began phasing out MSSS, which was replaced by the Distributed Mission Support System (DMSS). Like MSSS, it was a computer network with the mission-dedicated computers added into its general infrastructure.[107] ESOC installed two SUN workstations in 1987. One was a SUN 3/50 with four megabytes of main memory and the other was a SUN 3/160 with eight megabytes of main memory plus 141 megabytes on an internal disk. These were so successful that by 1990 ESOC had installed seventy SUN workstations. These new workstations did pose some difficulties for the electrical installation, because their hardware requirements differed from the previous workstations."[108]

Currently, the Operations Control Centre consists of two buildings, including the Main Control Room, the computer complex, various dedicated control rooms, and offices for the operations directorate. ESOC as a whole had grown from the original two thousand square meters to almost thirty thousand square meters in the 1990s.[109] ESOC consisted of a series of buildings that, for the most part, focused on spacecraft operations for ESA. The buildings housed administration, security, a cafeteria, a library, a myriad of offices, a computer complex, and the control rooms that are the focus of this study. Generally, the workforce consisted of 15 percent Flight Dynamics personnel, 19 percent engineers, 33 percent operations personnel, and the final third personnel for communications.[110]

The OCC, which comprised buildings D and E, remained the heart of ESOC. The controllers in the OCC provided scientists and experimenters with data from all instruments, information on orbit and attitude, relative distances and times, and any other critical information arriving from spacecraft. Those controllers also ensured the quality, quantity, and availability of data for all users.[111] ESA remained mindful that operations included not only hardware and software, but also procedures and personnel.[112]

The Main Control Room (MCR), in Building E, always had been the hub of the Operations Control Centre. Like JPL's Operations Room, it served as the central control center for critical aspects of missions, most notably LEOP. Despite its importance, it was only manned for a limited amount of time, because the majority of the time missions were running ordinary operations.

During the noncritical majority of mission time, spacecraft were controlled in dedicated control rooms (DCRs). Originally, when only a handful of missions were flying at any one time, each had its own DCR. More recently, with the ever-increasing number of spacecraft to be controlled, DCRs had become control areas for families of missions, rather than single missions. Thus, a single DCR may have housed facilities for earth observation, deep space, or Mars missions. ESA had learned to use the same or similar software and hardware across families of missions, thus allowing this marriage to work more smoothly. By using the same systems for multiple missions, it also limited costs and training needed to run the missions. In many ways, ESA's limited budget and resources, especially when compared to NASA's, had forced it to work more efficiently and, perhaps, more intelligently.

The OCC included an area for Flight Dynamics. This area specialized in satellite navigation, including orbit and trajectory analysis.[113] Standardization across projects was a key element for efficient work in flight dynamics. Personnel also ensured that their data were clear and unambiguous for whoever might use them. The Science Mission Support Section of Flight Dynamics handled testing and validation, mission analysis, and earth observation calculations, among other things. This section was further separated into four teams: attitude, commands, maneuvers determination, and orbits determination. The Navigation Office was named the Global Navigation Satellite Systems (GNSS) and could be used for a variety of products and tools.[114] Flight Dynamics included a Delta differential technician who can determine the angle of the spacecraft. This capability was critical for navigation and for observations. This technology was borrowed from NASA's JPL.[115] Flight Dynamics at ESOC currently includes about seventy members, which contrasts strikingly with the two hundred to three hundred at JPL.[116]

The OCC also included spacecraft simulators, which had been described as the "most important tool for validation."[117] Engineers used the simulators to prepare for any contingency, because trial and error was not allowed in commercial or professional ventures.[118] These simulators not only tested the hardware and the controllers, but also helped to emphasize the importance of teamwork rather than the individual.[119] Although early simulations were limited due to the available software and hardware, by 1977 and the GEOS mission, simulators could produce telemetry and accept telecommands from controllers to enhance the effectiveness of the training exercises.[120]

The Computer Control Centre (CCC), located within building D of the OCC, contains the mainframes for the MCR and the various DCRs. It currently uses SUN systems, and is transferring from Linux software to Solaris. The CCC used off-the-shelf hardware, which was upgraded and updated as much as the budget allowed. Unlike JPL's computer center, which housed a collection of hardware from across the decades of its existence, ESOC updated all of its hardware and did not hold on to outdated equipment. Controllers in CCC worked extended shifts from 6:00 a.m. to 10:00 p.m., so it was not manned continuously.[121]

The main control room for the ESTRACK ground network could be found in the ESTRACK Control Centre (ECC) of building E. The ECC typically consisted of three positions, a shift coordinator and two engineers. Shifts traded off throughout the day to maintain a presence around the clock. Because ESTRACK was automated, these engineers paid particular attention to any errors or problems. If necessary, they could call upon emergency personnel to fix any on-site problems within the network. ESTRACK also relied on standardized hardware and software throughout the network for maximum work output.[122]

One other area of ESOC deserves mention. Building H housed a ground station reference facility. This area could check the link between ESOC and spacecraft two years before launch. This allowed controllers to debug any problems in the connection. It served as a simulator for ESTRACK that could be vital to diagnosing problems before they were too far removed from the ground to be fixed.[123]

The OCC published a series of flight control procedures with instructions for normal and emergency procedures for each program. The Mission Implementation Plan served as a bible for each program and included virtually every detail needed for that mission.[124] The Flight Operations Plan (FOP) included the sequences, in detail, for various phases of each mission.[125] The FOP especially focused on all of the activities before and during LEOP.[126] The Satellite Data Operations Handbook organized information on how telemetry words and commands should be processed and displayed for the controllers. Finally, the Flight Dynamics Launch Support Document included vital information on the use of flight dynamics software during LEOP.[127]

The OCC, like JPL's SFOF, was a dynamic workplace. Not only was technology constantly upgraded, but the rooms also changed to house various missions. This flexibility has proven vital to ESOC's existence, and will serve it well in the future.

MAIN CONTROL ROOM

The European Space Operations Centre has served as the central control center for the majority of European robotic missions. By 2010, ESOC functioned as a control center, in some capacity, for sixty ESA missions and fifty other missions.[128] The ground network station in Spain has operated a few missions. The European Space Research Institute (ESRIN) houses the control room for earth observation missions. Some missions have had control rooms in other centers, including the Goddard Space Flight Center and JPL. Manned missions are usually run through NASA's JSC or the European Astronaut Centre (EAC) in

Cologne, Germany. ESOC employees also have worked at JSC for a few missions, including Spacelab and Eureka.[129]

Regardless of the location of their control center, the Main Control Room (MCR) of ESOC serves as the center for critical aspects, most especially the LEOP, for virtually all European missions. In this way, it resembles JPL's Operations Room more than JSC's control rooms. The MCR has even been used for rescue operations for various non-ESA spacecraft. Indeed, the success of their rescue efforts has made ESOC a go-to stop for many commercial endeavors. The December 1998 agreement known as the "Resolution on the Distribution of Tasks and Cooperation between ESA and National Flight Operations Control Centres and Facilities" came as a major coup for ESOC. It stated that only ESOC shall serve as the main control center for all future ESA missions, with the exceptions of the Columbus laboratory for the International Space Station and the Automated Transfer Vehicle (ATV).[130] The MCR has supported launch and early operations for Indian, French, German, and Italian telecommunication satellites, American and Indian meteorological satellites, Japanese earth observation satellites, and many more.[131] ESOC actively maintained healthy relationships with other spacecraft operators, even during the tumultuous Cold War years.

While the majority of commands originate from ESOC, certain routine or emergency commands can be sent directly from the various ground network stations, depending on the established rules for that spacecraft.[132] The rules for each flight were written in a mission handbook that detailed procedures for nominal mission events as well as certain anomalies. In fact, each of the control centers had similar guidelines for each of their missions. At JSC, for instance, the flight control team wrote a book called *The Flight Mission Rules*.[133] These mission rule books served as the bible for each mission. They allowed controllers to focus on more potentially pressing needs. It also sought to keep possible human factors from disturbing nominal work.

The layout of the Main Control Room has remained relatively constant over its decades of use, especially since *Helios-2*, which launched on 15 January 1976.[134] The room has three rows of consoles. The first houses consoles for the majority of the controllers. Eight members of

Main control room at ESOC, 2012. Courtesy of ESA/J. Mai, http://www.esa.int/spaceinimages/Images/2012/06/Main_Control_Room_at_ESA_s_Space_Operations_Centre.

the flight control team man the front row, with three computer screens for each station. The second row, though smaller, includes a console for the spacecraft operations manager. The final row consists of two separate sections of consoles for various liaisons and other staff that need to be in the room during the critical parts of the missions. The left-side console includes stations for the software coordinator, on the left, and the ground operations manager, on the right. The console on the right also has two stations. The flight operations director has a station on the left and the project representative, who maintains contact with the Project Support Room with representatives from industry, sits on the right.[135] With shift workers included in the tabulation, there are roughly forty operations positions available in the MCR.[136]

The MCR has always consisted of consoles, which, ESA argues, aid in tidiness and discipline.[137] Despite this relative constancy, the MCR must maintain a certain amount of flexibility due to the ever-changing needs of the different various projects. The MCR depends on high reliability in a real-time environment, quick system response, guaranteed availability of data, and clarity of information in order to function properly and efficiently.[138]

One major change to the room concerned Flight Dynamics. A glass wall originally separated the Flight Dynamics controllers on the left-hand side of the room. ESOC eliminated this wall to free up more space for engineers for multiple missions.[139] As a consequence, ESOC built a separate room for Flight Dynamics in the OCC.

Like JSC and JPL, the front of the room includes plotting maps and video screens. Below the screens are a series of clocks showing the local times for the stations of ESTRACK. Those clocks reaffirm the importance of ESTRACK to the ESOC mission. A list of missions controlled from the MCR, called the Spacecraft Date Launcher, hangs along the top of the walls on the right-hand side and the back wall. While it may be a less visual or symbolic remembrance than JSC's mission patches, it serves a similar purpose. The impressive list also serves as a visual reminder when ESOC may try to sell the MCR to potential space endeavor clients.

Like the previously examined control rooms, the MCR relies on low lighting so the controllers can concentrate on their individual displays. The MCR walls are dark in color. In that way, the room resembles the Operations Room of JPL more closely than the mission control rooms of JSC. At one point, designers planned on replacing the paneling with lighter-colored walls. Controllers balked at this idea. When asked why, Wolfgang Wimmer, lead flight operations director (FOD) for fifteen missions, explained that the dark walls and directional lighting allowed him and others to focus on their consoles and on the information directly in front of them. Lighter walls might have distracted the controllers when they could least afford it.[140]

As an example of the constant upgrading to the room, ESOC installed new workstations in 1990 with more user-friendly SUN equipment.[141] ESOC refurbished and upgraded the MCR again in July 1994.[142] The original layout remained largely intact due to the wishes of the controllers themselves.[143] Again, in this way the structure of the MCR remained constant, like JSC's MOCRs and FCRs but unlike JPL's Operations Room.

In 1988, the MCR began to use a new common software system compatible with any spacecraft, known as the Spacecraft Control and Operations System (SCOS).[144] Among other things, the SCOS brought major improvements, including dedicated hardware configurations for each mission, reutilization of common software across programs, support for telemetry and telecommands, and the use of commercial management systems.[145] This system received an overhaul in 1997 with the new-generation Mission Control System, SCOS-II.[146] The change came for a number of reasons, including financial, functional, strategic, and greater flexibility. It was described as better for both "normal" and "critical" operations.[147] Only two years later, ESOC upgraded it again, to SCOS-2000. In 1999, a fear grew among corporations around the world that, when the year changed to 2000, computers using only two digits for the date would revert back to 1900, thereby causing unknown damage to their systems. SCOS-2000 served as ESOC's answer, replacing non-Y2K-compliant systems.[148] ESOC configured SCOS-2000 to be used in both testing and operations for spacecraft.[149]

At the same time, they upgraded the various ground-station computer systems of ESTRACK so that they could be fully automated from ESOC during routine operations.[150] Full automation brings a number of benefits. The systems can only perform a small number of duties independently, which reduces the risk of human error during these operations. Perhaps most importantly, it reduces the needed workforce and saves money. While emergency responders are required to stay within a set distance of each ground station if need arises, a limited number of individuals can oversee the majority of the work in a more central location. Full automation has also changed the dynamic of shared ideas between ESA and NASA. Those in charge of NASA's Deep Space Network recently have begun to investigate automation for their network. Perhaps not surprisingly, they have examined ESTRACK closely for ideas and information on the process.[151] During an extended downtime for LEOP in 2011, the MCR underwent a major overhaul, including new consoles and a new visualized computer system. The updated MCR was prepared for a resurgence in operations in 2012.[152]

While computers and other equipment play a large role in missions, humans are still ultimately responsible for the success of the missions.[153]

Each mission begins with a ground segment manager (GSM), nominated to serve as the lead.[154] The ground segment manager serves in a similar role as the mission managers of JPL, though specifically for the ground segments of the mission. He or she must prepare and procure the necessary resources for the mission.[155] The GSM must also maintain the ground segment within the proposed schedule and budget. The GSM acts as an interface between the ground segment and the project, working especially closely with the project manager. In certain cases, heads of divisions for mission families work as GSMs for multiple missions. The GSM further establishes operations concepts and facilities as well as directing operations during LEOP and certain routine phases.[156]

The MCR gains control as soon as a spacecraft separates from its launch vehicle, roughly ten to fifteen minutes after liftoff, to the end of early phase operations.[157] Satellite LEOP is conducted by the flight control team (FCT) under the flight operations director (FOD). The FOD is not directly involved in mission preparation, so he usually joins later in the process, between six and nine months before launch, which contrasts strikingly with his counterpart at JSC, who joins the mission four years prior to launch.[158] The FOD must have a vast amount of experience in space operations as well as a familiarity with the ESOC operations environment.[159] The FOD relies on a consensus from the other controllers to make major decisions.[160] The FCT consists mostly of specialists in spacecraft operations and flight dynamics, including the spacecraft operations manager, the ground operations manager, and the flight dynamics coordinator.[161] FCTs may remain relatively constant over a short period of time or between similar missions, depending on each launch specification. A core team of approximately thirty controllers makes up the current FCT.[162] The FCT usually includes three or four on-call engineers, two or three on-call analysts, and six spacecraft controllers working shifts.[163] The LEOP team has always consisted of two shifts, due to a lack of funds for a third shift. They therefore must work up to eighteen hours under a lot of tension. Understandably, controllers may become overstressed, which leads to problems. Professional pride, however, reminds the controllers to maintain their focus and continue to work through the long shifts.[164]

The FOD serves as close a role to a flight director as ESOC has.

A FOD has overall control of a mission, but many of the commands are sent without direct approval. The FOD is also remarkably different from the flight director, because instead of making decisions unilaterally based on the given information, the FOD generally makes an assessment based on a consensus from the other controllers.[165] Also, due to the nature of ESOC operations, FODs often work on multiple missions simultaneously.

During the critical LEOP time, the MCR tests the spacecraft equipment to make sure it will function as needed. After the initial phase is complete, the MCR hands control over to the respective Dedicated Control Room (DCR), and it will only get involved again if needed for any other critical aspects of the mission. The DCR team can consist of members of the FCT.[166] Although the number of hours of use for the MCR is minor compared to the DCRs, they occur during the most important stages of the missions, therefore making it an indispensable part of spacecraft activities at ESOC. As Howard Nye, a former spacecraft operations manager, stated, the quality of and access to the end product of missions relies on the effectiveness of mission operations at the MCR to recover from any anomalies during the mission. Each mission is considered terminated at the moment the ground cannot remain in contact, regardless of the fate of the spacecraft.[167]

ESOC has repeatedly changed the dedicated control rooms over the course of their use. DCRs originally controlled one mission, hence the word "dedicated" in their title. They also originally consisted of consoles, looking similar to the Main Control Room. More recently, the DCRs, as previously explained, have become control rooms for families of missions. These families of missions allow for maximum synergy.[168] Recently, ESOC also has been replacing the consoles with everyday office equipment, such as tables and desks, to aid in flexibility. The increased flexibility also has been increasingly important during the transition to control rooms for mission families. "Dedicated" control room may be a misnomer now, but the terminology remains as a link to the past.

Controllers in the DCRs keep constant vigil over their respective programs. Unlike the MCR, which is staffed only during critical parts of missions, DCR controllers, especially for observatory satellite missions, remain on post virtually all the time.[169] These controllers exhibit a "calm efficiency" during any crisis.[170]

The Meteosat Operations Control Centre (MOCC) is just one example of the many DCRs. In 1977, *METEOSAT-1* became the first European weather forecast satellite. The MOCC served as the central facility for meteorological data.[171] Virtually all weather forecasts in Europe for the next few decades came from *METEOSAT* information. When Meteosat moved out in the mid-1990s, ESOC partially refurbished the MOCC for potential International Space Station (ISS) operations.[172]

Another DCR controlled the GEOS missions of the 1970s. The consoles included alphanumeric displays that displayed real-time data, a processing system, recorders for telemetry data as a function of time, graphical displays for data from the spacecraft or experiments, and keyboards to input information and interact with the computers. Scientists also had full control of any experiment.[173]

In 1994, ESOC completed construction work on a variety of DCRs for individual missions in preparation for the Cluster mission.[174] These DCRs, like the others, contained individual differences in accordance with the needs of the particular program. It is interesting to note that the DCRs for Cluster as well as earth observations had windows, a stark change from most of the interior-roomed control rooms.[175] Nine years later ESOC added new DCRs for the SMART, Mars Express, and Rosetta programs.[176]

The Planetary Missions Control Room is one of the oldest DCRs still in use. This DCR uses consoles and so has not been upgraded to the office furniture used in most DCRs today. This DCR is the primary facility for many deep space missions.[177]

The Venus Express DCR, constructed for the 2005 mission, reused various ground segment elements and operations from previous programs, which reduced both the cost and risk factors. This particular DCR has been called the Venus Express Mission Operations Centre, or VMOC. It includes a Mission Control System, a Data Disposition System to acquire and store data, a Mission Planning System, an independent Flight Dynamics System, and a Spacecraft Simulator. The VMOC also includes the Venus Express Science Operations Centre (VSOC) for use during critical mission objectives.[178]

The operations in ESOC's Main Control Room resemble JPL's Operations Area more than JSC's control rooms. Like JPL, the MCR is generally only staffed during LEOP events. The majority of the controllers staff the DCRs, which are located throughout the OCC. Finally, neither ESOC nor JPL have a figure with as much authority as JSC's flight directors, instead relying on handbooks to provide commands for nominal control.

CONTROL ROOM WORK

ESOC controllers view spacecraft as machines built to produce results, and they recognize that mission products must be optimized. As a result, they sometimes decide between quantity requirements and quality requirements. In some ways, controllers see themselves as a service entity, providing a product to a customer. Part of that service is knowing the time sensitivity of their different products, or data. Keeping this in mind, before the mission they calculate the maximum data output from their spacecraft and aim for at least 98.5 percent output. They even plan to save fuel in order to extend the life of the spacecraft, if possible and if required.[179]

Controllers in the OCC can be grouped into a few different positions. Network controllers make sure that those working in the DCRs maintain contact with anyone outside the DCR. Spacecraft controllers are those continuously manning the DCRs. Orbit or attitude controllers ensure that the latest information on the status of the spacecraft is available to anyone who might need it.[180]

There are two classes of controllers at ESOC. Engineers have extensive higher-level education and usually have a master's degree in a certain specialty. Engineers mostly work regular business hours, but they can be called in if an emergency occurs. They typically only work specific missions pertaining to their specialty. Operators and analysts, on the other hand, may have a technical degree. They work shifts and man the consoles twenty-four hours per day. These controllers follow strict procedures, calling an engineer if something goes awry.[181] In other words, engineers are specialists, whereas operators and analysts are workmen. Engineers are almost always staff, whereas the operators can be staff or, more likely, contractors. This can sometimes lead to further segmentation between the engineers and operators.[182] The staff-to-contractor ratio usually remains around one to three. Mission operations currently includes eighty staff members and 230 contractors.[183]

One of the major differences between ESOC and JSC or JPL lies in the number of staff. ESOC consistently comments on the difference, which is sometimes by powers of ten, in team sizes. The main reason for this difference is the historically vastly different budgets for the space agencies, which depends largely on their roles in the Cold War. A controller team for a mission may consist of three engineers and a few technicians, while the flight dynamics team may include up to ten people. Due to the staff differences, ESOC controllers often are required to work multiple missions at the same time. Howard Nye, for instance, served as an FOD for Giotto, the spacecraft operations manager for Hipparcos, and the spacecraft operations manager for Iso between 1989 and 1993.[184]

The challenge of the work often serves as a natural way to distinguish between those who are meant to work the difficult job of controller and those who are not. Those who are committed are driven by the continual changes induced by new missions and new technologies. The job naturally attracts the more adventurous. Given the multinational staff, ESA employees must also be open to different cultures. The few who cannot let go of their home nationality also usually do not last long. In fact, a feeling of statelessness can bring the ESA staff together.[185]

ESOC has referred to simulations as the "most important tool for validation" of spacecraft.[186] It took some time for ESOC to produce effective simulations. Simulators during the first decade of control at ESOC were limited, due to technological restrictions. In 1977, new simulators began to produce telemetry and accept telecommands for more complete and accurate simulations. These came online for the GEOS missions.[187]

ESOC also produced videos for the initial training of controllers. Those videos were not used to a great extent because they were not as effective as training through simulations. ESOC also focused on simulations for training to emphasize the team rather than the individual. In the control room, controllers must always be aware of how their actions relate to the entire team. Controllers acting individually could potentially compromise the rest of the team.[188]

ESOC is perhaps the most accomplished control room for rescuing missions. While ESOC has never lost a mission, it has rescued a number of missions for other organizations.[189] The first rescue mission for ESOC occurred with *TD-1A*, which launched on 12 March 1972 as an astronomy satellite. By the end of May, all the tape recorders on the satellite had failed. As a result, for about two years, ESOC personnel were required to work in seven mobile stations located around the world, with an additional seventeen ground stations, in order to collect as much information as possible from the direct signal downlinks.[190]

A few years later, problems during the April 1977 launch of the *GEOS-1* satellite to investigate the magnetic environment in space placed the satellite in a bad orbit. Over five days of hard work, the controllers fixed the orbit to the best of their abilities, which allowed the spacecraft to continue its scientific mission.[191]

One of the most heralded efforts to come from ESOC was the Giotto program. It was originally designed and built for a close encounter with Haley's Comet in 1986. Symbolizing the potential success of increased international efforts in space, Giotto ESA controllers worked with those of NASA and Russia to locate the comet.[192] Information travelled from the Soviet Vega through NASA's DSN to ESOC, in true international cooperation.[193] After a successful rendezvous, the engineers placed it in hibernation. In 1990, controllers reactivated the spacecraft after four years of hibernation in order to attempt another rendezvous, this time with Comet Grigg-Skjellerup in 1992. This marked the first time that a spacecraft had been placed in hibernation and reactivated years later, an important accomplishment with major implications for future deep space missions. This also marked another instance of international cooperation, since ESOC controllers were able to use NASA's DSN station in Madrid to contact *Giotto*.[194] *Giotto* was a major public relations success story for ESOC.[195]

The Hipparcos mission, focusing on star measurements, remains one of the most important recoveries in spaceflight history.[196] Launched in August 1989, *Hipparcos* experienced the failure of an apogee boost motor shortly after liftoff, leaving the spacecraft stranded in its transfer orbit. Controllers worked tirelessly to implement a revised mission plan. After more than three years in orbit, *Hipparcos* completed all of its stated objectives.[197]

Another satellite launched in 1989, *Olympus*, required a rescue mission. This telecommunications satellite was controlled out of Fucino, Italy. During its mission, the satellite experienced major onboard failures and even lost 50 percent of its solar powers. It later began to tumble out of control after switching to safe mode. In 1991, ESOC controllers were called on to thaw its frozen batteries and propellant and regain control. After their success, *Olympus* continued in service until 1993.[198]

ESOC remained the most successful rescue control center in the world in the 1990s. In 1998 controllers lost contact with *SOHO*, a satellite that they had launched in 1995. ESA and NASA engineers working together were able to approximate the location of the satellite and pick up its signal for two to ten seconds. After three months of efforts, the controllers successfully reoriented *SOHO* to the sun and regained contact. Another satellite launched in 1995, *ERS-2*, began to exhibit gyroscope failures in 1997. By February 2000, ESOC controllers had established a new control mode to fix the problem. *Huygens*, which had launched in October 1997, experienced problems with its radio link in February 2000. Again, controllers fixed the problem in September 2003. During this time, ESOC continued to help with the recovery of satellites for other entities. For instance, controllers fixed the orbit of *Artemis*, a Telespazio satellite launched in July 2001. ESOC also helped rescue the *ETS-VII* and *COMETS* satellites for NASDA, the Japanese space agency.[199] *ETS-VII*, a satellite launched in 1997 to test robotic rendezvous, began to spin on its axis, and in roughly five hours, ESOC was able to establish contact, uplink a software patch, and reorient the satellite. *COMETS*, a broadcast communications satellite, endured a partial failure during launch in 1998 and could not reach its intended apogee. In order to fix this problem, ESOC supported NASDA with emergency assistance through ESTRACK using infrastructure from *ETS-VII*.[200]

Because of these rescues and the success of ESA missions in general, ESOC controllers have been lured away from the agency by various private companies in need of well-trained controllers, who can be difficult to find. Higher salaries in the private sector can be especially enticing.[201] This problem plagues government organizations across the world.

*　　*　　*

ESOC has various similarities to the main control centers of NASA. The Main Control Room has remained essentially the same throughout its history, more like JSC's than JPL's. Operations, however, resemble JPL more than JSC in that the main room is used only for critical aspects of missions. The majority of control work occurs in other dedicated control rooms.

Like JPL, ESOC has no individual with the power of a flight director. Also, ESOC and JPL have become adept at monitoring multiple missions simultaneously. In most aspects, it seems as though JPL and ESOC are more similar in the nature of their robotic missions, versus the constant vigilance needed for human spaceflight out of JSC.

ESOC does contrast strikingly with the American centers in one important regard: budget. Throughout its history ESA has managed to successfully complete impressive spaceflight missions with a budget much smaller than NASA's. This has forced ESA and its centers to focus on more limited goals. ESOC has also maintained a much smaller workforce, sometimes by powers of ten when compared to JPL and JSC. The smaller budget also helped ESOC lean more toward cooperation than competition, particularly during the high-budget NASA days of the Cold War. It has also forced ESA to automate aspects of its work, such as the ESTRACK network, earlier than JPL and JSC. Due to that change, NASA now looks to ESA as an example of how to operate effectively with less means. Undoubtedly, ESOC provides a comparatively useful tool for the mission control centers of NASA.

4

INTERNATIONAL COOPERATION

Much of the emphasis in space history, particularly as the public perceives it, is placed on competition during the Cold War. That this competition fueled the Apollo program and the space race cannot be denied. Even the birth of NASA, where previously President Eisenhower had avoided creating a government space agency, came about due to the Soviet success with *Sputnik 1*. During the decades of competition, the space agencies did seek to work together under certain circumstances. As the Cold War dragged on, and as space budgets were steadily reduced, space agencies found it increasingly important to seek partners in completing their missions. This culminated in the largest peacetime international effort in world history: the International Space Station.

Cooperation and competition are not mutually exclusive. NASA and ESA, for instance, have worked together while also competing in space endeavors. Even during moments of cooperation, NASA tended to be more reluctant to share its information with its partners, fearing the loss of exclusive technical knowledge. NASA even had moments of cooperation with its fiercest competitor, the Soviet Union, during the Cold War. By spurring each other to grow, space agencies create better partners.[1]

This leads to the most important rule of international cooperation in space. Both countries must have a mutual interest in the project with potential benefits for both partners if it is going to succeed.[2] Any one-sided project will likely not move beyond a planning stage before at least one partner cries foul. Countries invest in space agencies expecting notable returns. The high price of spaceflight does not allow any country, regardless of size or economy, to fund ventures with no appreciable return.

The majority of the time, each partner must fund its own aspect of each mission. While there are some exceptions, a country rarely pays for another's activities. Again, this is in part due to the high price of spaceflight. It also ensures a good partnership. Each country can trust the other to satisfy fully its aspect of the mission.

Roy Gibson, former director-general of ESA, warned that national space agencies must continue to seek cooperation with each other. He feared that some only sought help when the size and scope of certain projects grew too large for one nation alone to finance. Even if singular national space agencies are profitable, they must continue to work together for the betterment of all humanity.[3]

As the most prominent face of the ground segment of space travel, mission control has often been at the forefront of international relations for the space agencies. As with so many other aspects of the control rooms, JPL and ESOC tend to be more similar in their acceptance of international cooperation, while JSC typically is more reluctant. While the majority of the discussion considers the broader involvement of the space agencies, this chapter will focus as much as possible on the role of the control rooms in international cooperation.

ESA

ESA is unusual among the early Western space agencies in that it was born not based on competition and "winning" the Cold War but, rather, based on cooperation. Thus, collaboration with other space agencies has always been primary for ESA. Former Director-General of ESA Hermann Bondi once remarked that international cooperation was both "essential" and "difficult," but any difficulties could be "overcome with sufficient effort and determination."[4] He further argued that international cooperation is not any less difficult than obtaining the technology needed for space travel, and thus it deserves the same attention and effort.[5] ESA is an unusual example of international cooperation because, by its very nature, it is an international organization. This section will discuss only interactions with space agencies other than NASA.

ESA's reliance on international cooperation was a natural outgrowth of post–World War II Europe. Stacia E. Zabusky discussed this concept in detail in *Launching Europe: An Ethnography of European Cooperation in Space Science*. In short, European countries surveyed the damage on the continent after two world wars and wanted to avoid such destruction again. As a result, they strove for a shared identity with the hope that commonalities would overcome any differences. Europeans also believed that they could reap more economic benefits from working together than they could as individual nations. In fact, none of the individual countries could compete with either the United States or the Soviet Union economically. Similarly, a unified Europe could negotiate with the United States and Soviet Union superpowers on a more equal footing than the individual countries could. These three main motivations made cooperation among European nations the standard approach for economics, politics, and society.

Advanced science and technology generally entails big budgets. European nations worked together to help defray the costs. With space as one of the largest postwar technological ventures, it made sense for Europeans to work together in a unified space program, since no single nation had the wealth or the infrastructure to maintain a large space venture. ESA could be considered a natural outcome not only of European politics and economics but also of European science and technology.[6]

Some have even argued that the Cold War between the United States and Soviet Union made it inevitable for European nations to work together in space. The perceived space race forced European countries to pursue an active space program if they wanted to be effective on the international scene.[7] A special commission in 1967, called the Causse Report after its chair, Jean-Pierre Causse, further enumerated the necessity of international cooperation. It stated, in part, that while ESA should remain independent, it should strive for close associations with the two superpowers, because ESA could not survive competing with them.[8] With these thoughts in mind, one can easily move on to the development of ESA working with outside space programs.

Since its inception, NASA has been ESA's closest international partner. Even when not collaborating on missions, the two have shared space technology and science. This special cooperative relationship receives full consideration later in this chapter.

While the extensive cooperation with NASA revealed ESA's loyalties during the Cold War, ESA did work with the Soviet Union as well. Europe's first astronauts, for instance, flew in Soviet spacecraft. This was due largely to the fact that the Soviets flew astronauts from varied countries well before NASA did. Foreign astronauts aboard Soviet spaceflights came from Afghanistan, Syria, Vietnam, Cuba, Mongolia, East Germany, France, and Austria.[9] The first non-Soviet European in space, Vladimir Remek, flew aboard *Soyuz 28* in 1978. NASA launches did not include a European astronaut until Ulf Merbold flew aboard *STS-9* in 1983. As late as 1989, ESA was continuing to work toward a formal agreement for cooperation in space with the Soviet Union.[10]

Some ESOC employees experienced minor difficulties working with Russians. The Soviet space agency had some different ways of accomplishing their goals. Some Europeans regarded the Russians as too limited in their work. The controllers and engineers did not have the freedom that those at ESA or NASA had. The language barrier could be extremely difficult as well.[11] The majority of European flight controllers could communicate in at least English and French, if not also German or Italian. Few spoke Russian. That just a small percentage of the Soviet controllers spoke any language other than Russian often caused tensions. Consequently, the lack of a common language across all space agencies has often restricted international cooperation.

ESOC is one of ESA's most valuable assets for cooperation with foreign space agencies. ESA has even published an ESOC Services Catalogue to be used to attract new customers. Among other things, it highlights the expertise of their controllers and the flexibility of the control center and ESTRACK. ESOC also emphasizes its perfect track record for European-launched spacecraft.[12]

The control room of ESOC has supported launch and early operations for numerous foreign space agencies. Through 2000, for instance, forty-nine of the ninety-seven missions controlled at ESOC were for external customers.[13] By 2010 that number had grown to nearly half of the approximately 120 missions controlled at ESOC.[14] These have included telecommunications satellites for the Indian, French, German, and Italian space agencies. ESOC has also supported a meteorological satellite maintained by the United States and India, as well as two earth observation satellites for Japan.[15] Paralleling the measured development of international cooperation on all matters of consequence, ESOC's track record has been particularly important for ESA's good reputation in the international community. There can be little doubt that ESA's unique position as a space program comprising members from various nations has aided its efforts to be a more cooperative space agency than the national programs of the United States and Russia.

NASA

While the public may perceive NASA as a national agency conducting its own projects without the help of outsiders, the space agency's charter contained numerous mentions of international cooperation. Seeking positive relationships with other space agencies has been a major goal for NASA since its inception. During the Cold War, NASA was an extension of national efforts to "win over" other countries during the conflict with the Soviet Union. This section includes just a few examples of work with foreign nations, excluding ESA.

When NASA was created under Public Law 85-568, the National Aeronautics and Space Act of 1958, one of the objectives clearly stated was cooperation with other nations for peaceful purposes. The act "emphasized openness and scientific objectives" regarding interactions with other countries. This is because the United States acknowledged that it could gain significantly from international cooperation. It had, therefore, been a major consideration for NASA since its inception. The political environment of the United States in particular, and the world as a whole, played an important role in how NASA approached international cooperation over the course of its existence.[16]

NASA guidelines for working with other space agencies dictated that JPL and JSC must follow rules when cooperating with foreign control centers. Cooperation occurs on a project-by-project basis and is neither open-ended nor nonrestrictive. The project must have mutual interests with clear scientific value for all parties. The technical agreements must be established before any political agreements are made. Each partner must take full financial responsibility for its own share of the project. Each partner must also provide full technical and managerial capabilities for its share of the work. Finally, the science from the project must be placed in the public domain.[17] In reality, these guidelines are practiced by all space agencies.

The United States had always invested in its space program far more than any other nation, with the possible exception of the Soviet Union. Since negotiations must be mutually beneficial, this led to many difficulties when attempting to cooperate with other space agencies. The gap could be substantial. For instance, in the mid-1960s, NASA's per capita expenditure exceeded $30. This figure decreased by half by the 1970s, but even then it far surpassed that of European nations, for instance, which spent on average about $1.50 per capita. This average never rose above $2.50 per capita for ESA.[18] Thus the size of NASA's programs and those of other space agencies differed greatly.

NASA was initially pushed to international cooperation largely due to the need for tracking sites worldwide for better coverage to control spacecraft.[19] Both the STADAN and the DSN utilized antennas at sites across the globe. These sites could only be built in countries with favorable relations with the United States, a requirement that caused a few problems. For instance, a site in Cuba had to be moved before the Bay of Pigs invasion in 1961. Another in the former state of Zanzibar was evacuated during a political uprising in 1964. In both cases, an unfavorable government forced NASA to rethink its antenna location. Thus, ground stations have long served as a means of diplomacy for the United States. The ground-station portion of the cooperation remains paramount, and both sides do their best to maintain compatible networks.[20]

The United States also viewed international cooperation as an important avenue for global markets. During the Cold War, the United States argued for a global free market economy and used numerous devices to gain advantage for their agenda in other countries. This played an especially important role in the communications and aerospace industries.[21] Cooperation has also been viewed as the "embodiment of peace,"[22] an especially important principle during the strain of the Cold War.

During the early Cold War, much of the technology flow originated in the United States and transferred to Western Europe and other countries. Recognizing this, the government wanted to limit technology transfer as much as possible, sharing only a fraction of that available. The initial guidelines for space science cooperation included scientific validity, mutual interest, specific rather than general spaceflight goals, the widest dissemination of results, and individual countries' taking responsibility for their own expenditures.[23] The scientific output of cooperative programs has been deemed so vital that any ventures without significant scientific gains are regarded as politically meaningless.[24] In many ways, space science was viewed as a benign transfer of ideas, and so it was more freely shared than others. Not everyone agreed with this sentiment, however. Some Americans did worry that working with Europeans in high technology could create stronger competition in the world market.[25] Given the bipolar nature of the Cold War, the United States did not want to make more enemies.

As an example of NASA relations with European space agencies, the Helios project, in the mid-1970s, was a joint project between JPL and the (West) German Space Agency to conduct solar research. The German Space Operations Center (GSOC) at Oberpfaffenhofen, near Munich, served as a primary control center for the launch phase and other aspects of the project. Thus, NASA gained greater cooperation with West Germany and more flexibility for the Deep Space Network.[26]

The relationship between NASA and the Soviet space agency remains a complex story of competition and cooperation. Between the launch of *Sputnik 1* and the culmination of the Apollo program, the competition aspect of the space race dominated the landscape. Both sides challenged each other for ever-greater accomplishments as a sign of political strength. Even JPL had to refocus its efforts to aid the Apollo program. Following the "victory" in space with the Apollo program, space budgets began to be reduced and the United States pursued a new foreign policy of détente. Both superpowers worked more diligently with other space agencies to complete their missions, even working together on a few projects.

A space-related agreement between the United States and the Soviet Union signed in 1972 allowed for the successful completion of, among other ventures, the Apollo-Soyuz Test Project (ASTP). ASTP culminated in the first docking between American and Soviet spacecraft. Most NASA employees viewed ASTP as a waste of time and resources, since it did not contribute anything new and was seen only as a political stunt. The agreement terminated in 1982, and due to contentious relations between the two superpowers, it was not renewed. Despite this, the DSN worked together with the Soviet space program to gather telemetry data from their two Vega balloons at Venus in 1985. The French space agency, CNES, also aided this international effort.[27]

During the preparations for ASTP, various members of both space agencies' control centers visited each other for discussions on the proper course of action for the mission. When the Soviet delegation visited Houston, JSC arranged and paid for their lodging and provided a fourteen-dollar per diem as well. The Soviets were also granted group hospitalization coverage in the event of an emergency. All transportation costs were paid for by the country sending the individuals.[28]

Interactions between NASA controllers and those of the Soviet space program could be difficult, even during their cooperative ASTP. At times there were issues of trust. Some NASA workers had the impression that the Soviet delegation expected them to be spies from the CIA. Interestingly, this rationale came from their expectation that some of the Soviets were KGB spies. They also felt as though they were being tested at first; but as interactions increased over time, that changed to a more mature working relationship.[29] Many of the NASA delegates have also commented that they feared that their rooms had been bugged, so they were careful about their conversations.[30] Despite their penchant for secrecy, many of the NASA controllers understood that the Russians were as complex as anyone else. In fact, they found out rather quickly that the Russians enjoyed partying and that vodka could bring out a more agreeable side.[31] As a testament to the increased level of cooperation and trust between the superpowers, Chuck Lewis, a flight director since the last Apollo missions, led the NASA flight control contingent in the Russian mission control room.[32]

One of the most difficult aspects of ASTP was the language barrier, both in space and between the control centers. A single word used by NASA could have a half dozen or more variations in Russian, and vice versa. The two sides sometimes found that they were arguing about a word when the concept they were trying to convey was actually the same. The language barrier also helped NASA employees realize how much jargon they routinely used and how difficult it could be for outsiders to understand them.[33]

Any potential cooperation with Russia, especially during the Cold War, had to keep political consequences in mind more so than relations with allies. Even when they cooperated, they remained each other's primary competitor in science and technology. Similar political constraints more recently have hindered greater cooperation between the United States and China, another emerging space power.[34]

As NASA began to turn to the shuttle, international cooperation became even more important. The method of collaboration changed as well. The growth in maturity and in the capabilities of ESA and other space agencies, the increasing cost of space missions, and the relative scarcity of funds for NASA all played a role in this transition.[35] The European Spacelab and Canada's robotic arm are prime examples of NASA's willingness to allow other organizations and countries to build key space components.[36] Cooperation between NASA and ESA, which will be discussed in the following section, grew more complex in the 1980s and 1990s as both agencies adapted and grew.[37]

Like JSC's human spaceflight program, JPL has worked more diligently in the new millennium toward international cooperation, with projects such as the Gravity Recovery and Climate Experiment (GRACE) with the German Space Agency and the Ocean Surface Topography Mission (OSTM) with the French Space Agency. Similar to ESOC, though on a smaller scale, JPL has served as a control room for launch operations for various other space programs. JPL recently has considered an international mission to Mars. As budgets are cut and the Cold War recedes further into the distance, NASA will only continue to rely on international cooperation to complete its missions.

NASA AND ESA

NASA always has been the closest non-European partner for ESA.[38] Likewise, ESA has been "NASA's foremost partner" in space.[39] ESA's main publication in February 1972 even showcased an open invitation from NASA for ESA staff to visit various centers in the United States.[40] The following year, ESA created an ESA Washington office at NASA's headquarters to foster "strong cooperative ties."[41] By 1991 the Washington office remained small, employing only four people.[42] Despite its size, it is still a crucial link between two of the world's largest space agencies.

From its foundation, NASA has recognized Europe as its primary, though not its exclusive, partner in space.[43] In fact, to demonstrate the specialness of the relationship, NASA launched ESA's first two satellites in 1968 for free as a "christening gift."[44] The United States gained more influence in Europe, largely due to the Soviet Union's self-imposed isolation, at least regarding technology transfer. Much of Europe viewed technology as a way to resurrect their economy and industry and thus placed a high value on advances in technology. They also wanted to avoid a technology gap as much as possible.[45] A close relationship with NASA, and the sharing of technology, played a key role in preventing or diminishing any such gap.

ESOC has primarily worked with JPL, because they both focus on robotic spacecraft. The cooperation between DSN and ESTRACK has provided further avenues for close cooperation. The Space Link Extension Services (SLES), established near the start of the new millennium, allows for JPL and ESOC to use each other's ground station networks more easily. This is a first-of-its-kind link of extraordinary international cooperation.[46] Many ESOC controllers and engineers have expressed that they have a great working relationship with those at JPL. Some have even worked together enough to consider each other friends.[47]

However, those professional friendships, naturally, never took precedence over their individual projects.

As with other examples of international cooperation, for each instance of NASA and ESA working together, the parties must approve a memorandum of understanding (MOU) that explicitly states how they will work together and what they will share. More recently they have used more generic and overarching technical assistance agreements (TAA) to bypass the need for a specific MOU.[48] TAAs work much like NASA's Discovery program in that the federal government approved a set of regulations and NASA was allowed to make specific decisions within those guidelines.

MOUs present an interesting and important distinction between NASA and ESA. NASA is a national agency under the jurisdiction of the government of the United States. ESA, on the other hand, is an international organization under the jurisdiction of international law.[49] The ESA Convention grants the space agency the power to create and sustain treaties with other governments, representatives, or agencies. In effect, the director-general initiated any treaty, and it required unanimous approval by the members of the council. This leaves ESA and NASA, and in fact any other national space agency, on unequal grounds. NASA cannot sign a treaty, a formal, binding document, on its own. MOUs and other arrangements are thus necessary.

Because MOUs hold a less formal status than treaties, there can be undesirable, even damaging, consequences to using them. If ESA and NASA could sign a formal treaty to conduct a program, the two partners would be required to complete that agreement regardless of any unforeseen complications. Because the two space agencies can only conduct informal agreements, however, MOUs are not as strictly binding. NASA regards MOUs as executive agreements that do not require Senate approval. NASA has even included clauses stating that each partner will make their "best effort" to complete the project, thereby weakening any binding characteristics.[50]

The International Solar-Polar Mission (ISPM) provides a significant example of the vagaries and misunderstandings that can come about as a result of the lack of formal treaties. NASA and ESA signed an MOU in 1979 for a mission that entailed each agency's constructing a satellite to be placed in solar orbit over its poles. ESA obtained the funds for its satellite almost immediately. NASA waited until 1981, at which point budget cuts made it impossible for the space agency to pursue its satellite. NASA declined to fulfill its role in ISPM on the grounds of lack of funds and "best efforts," as detailed in the MOU. ESA and other European officials protested that they had a formal agreement to complete the mission. NASA countered that each side signed the MOU with full knowledge of the conditional clause. In the end, ESA continued its project under the new name Ulysses and proceeded to conduct future projects with NASA with a degree of distrust.[51] Despite this issue, JPL housed a Dedicated Control Room (DCR) for the Ulysses mission, staffed by ESA employees.[52]

Like their relationships with the Soviet Union, though on a different scale, ESA and NASA cooperated with and competed with each other simultaneously.[53] That being said, most ESA employees view their relationship with NASA, and vice versa, as a good and cooperative partnership.[54] It remains an "intensive collaboration."[55]

Since the inception of ESA, English has served as the official language, despite it being the primary language for only approximately 15 percent of its employees.[56] There were a few reasons for the choice. English was recognized as the major language of international commerce. It did not favor any of the major continental member states. Perhaps most important, English made interactions with NASA, particularly in terms of the control centers, more fluid. French served as another official language, so ESA staff members were expected to speak both languages. Because those languages were not the native tongues for a large portion of the ESA staff, they were required to learn a more technical and less conversational common version of the languages. This challenged the native speakers. Irish or British staff members, for instance, had to learn to avoid using cultural idioms that others might not understand.[57]

The politics of space was especially important during the Cold War. Much of the United States government's actions during the Cold War should be regarded in light of how it affected its relationship with the Soviet Union. As such, the government considered it "political goodwill" if NASA could be shown publicly cooperating with European efforts in space. These actions resonated domestically as well as internationally. Americans were comforted knowing that the alliance with Western European nations was growing. Western Europeans continued to view the United States as a closer and more viable ally than the Soviet Union. The NASA Task Force of 1987 agreed that cooperation was motivated primarily by foreign policy decisions.[58] In short, the United States government recognized that NASA served as an essential extension of Cold War foreign policies toward other nations, and especially toward Europe.

Much of the cooperation between NASA and ESA began in 1965. Arnold W. Frutkin, the NASA assistant administrator for international affairs, and Pierre Auger of ESA began informal talks regarding NASA launching ESA satellites with compensation. They agreed that ESA would give NASA any information about spacecraft performance generated by the launch. The following year, NASA approved an MOU with ESA stating that NASA would provide reimbursable launch services. Each launch required a separate contract and a designated project manager to coordinate all interactions.[59]

After the Apollo program, a number of changes affected international cooperation, particularly between the United States and Europe. The United States government greatly reduced NASA's budget. NASA encouraged the Europeans to take a more substantial role in space, which coincided with growth at ESA. The United States also acknowledged a rising concern about sharing potentially sensitive technology with other nations.[60]

In a prime example of cooperation leading to competition, and in an effort to break its reliance on other space agencies to launch its spacecraft, ESA embarked on perhaps its most ambitious solo project in 1973: *Ariane*, an expendable launch vehicle. The first successful launch on 24 December 1979 broke NASA's monopoly on commercial launch services in the West. The five generations of *Ariane* have allowed ESA to compete with NASA for other organizations' launches.[61]

Perhaps the most impressive robotic spaceflight collaboration between NASA and ESA came with *Cassini-Huygens*. Following more than a decade of development, *Cassini* launched on 15 October 1997 at Cape

Canaveral. The satellite and probe arrived at Saturn on 1 July 2004. On 25 December 2004, the *Huygens* probe, developed by ESA, separated from NASA's *Cassini* to enter the atmosphere of Saturn's moon Titan. The two space agencies then controlled their respective spacecraft, with JPL overseeing the *Cassini* orbiter and ESOC monitoring the *Huygens* probe. ESOC had also sent commands to the probe from its specially developed Huygens Probe Operations Centre (HPOC) once every few weeks during the seven years between launch and arrival. *Huygens* successfully landed on the surface of Titan on 14 January 2005 and continued to send data and images for ninety minutes. This remains the only landing on an outer solar system planet. *Cassini* is scheduled to remain in orbit until 2017.[62]

Spacelab is perhaps the most important collaboration between ESA and NASA for human spaceflight prior to the ISS. In many ways Spacelab, like many other projects from this time, grew out of NASA's decreasing budget. In the 1970s and 1980s, NASA's human spaceflight program spent so much of its budget on the development and launching of the space shuttle that it had to find partners to develop the scientific projects to be pursued by shuttle crews in space. ESA and NASA signed an MOU in 1973 in which ESA agreed to build a reusable space laboratory to be launched multiple times by the shuttle. While ESA built the actual laboratory, other space agencies, including those of Germany and Japan, eventually equipped Spacelab with experiment racks and platforms. Spacelab launched twenty-two times between 1983 and 1998, making it the primary scientific platform for the shuttle program.[63]

The United States government reduced post–Cold War NASA's budget even more. Foreign policy and economics emphasized international cooperation. NASA was also forced to focus on smaller, more affordable projects and a quicker turnaround of output and results.[64] This led to the "faster, better, cheaper" initiative at NASA. Many of the ESOC controllers reviled NASA's new mission statement, thinking it amateurish and an insult to spaceflight professionals. They have pointed to the numerous failures in low-cost projects, and their own flawless performance with their medium-cost missions, as proof.[65] Changes in policy, however, did not hinder collaboration.

NASA continued to recognize the need for cooperation between the United States and Europe. Daniel Goldin, the NASA administrator who advocated "faster, better, cheaper," spoke of a "common destiny in space" and said that the relationship with ESA was "vital to the US space program." He further stated that they "must complement one another," and that they must strive for "a bold vision, a common set of objectives that will allow (them) to work in space together, not separately."[66] Not surprisingly, he focused on ideas of cooperation while downplaying potential competition. Increasingly, NASA realized that accomplishing their goals with a depleted budget meant depending on international partners in space.

Following the terrorist attacks of 11 September 2001, the free exchange of information between ESA and NASA came to a halt due to the restrictions placed on international information exchange by the newly founded Department of Homeland Security and the USA PATRIOT Act.[67] ESA and NASA came to an agreement in 2007 for mutual help to circumvent bureaucratic approval. A technical agreement between ESOC and JPL was signed by the United States State Department.[68] Despite these advances, sharing information remains largely one-sided, now from ESA to NASA, with more limitations on how freely NASA may communicate.

One final example of the dynamic cooperation and competition between NASA and ESA was Giotto. When ESOC first decided to expand into interplanetary missions with Giotto in 1985, it benefited from JPL's early success with such missions. ESOC could learn from their experience, especially with the difficulties dealing with deep space communications. More recently, JPL has come back to gain information about ESOC's Ground Station Network. JPL is especially interested in ESOC's ability to operate its ESTRACK stations remotely, which JPL's DSN cannot do. ESOC's expertise in automation reduces the chances for error under nominal conditions by eliminating human error. In order to address problems or deviations, each station is required to have at least one controller on call and within two hours of the station at all times.[69] As Manfred Warhaut, an ESOC veteran, describes it, in many ways the "cross-fertilization" between NASA and ESOC has been beneficial.[70]

INTERNATIONAL SPACE STATION

International cooperation has grown increasingly important for space agencies coping with shrinking budgets and heightened public expectations. No other project highlights this change more than the International Space Station (ISS). The ISS is the largest peacetime international endeavor in human history. It is also the most impressive example of multinational collaboration in space history. For this reason, it will serve as the primary case study for international cooperation in space.

The ISS story begins with NASA's proposed space station Freedom. In 1984 President Ronald Reagan announced his wish for a permanent presence in space with a new space station. Over the next year, NASA engineers began to design a station with multiple modules for experiments, power, and habitation. As the project, and hence the need for funds, grew, NASA reached out to other space agencies to contribute to an international effort.

NASA approached ESA, Canada, and Japan in 1985 about joining the space station effort. This can be viewed as not only essential due to budget constraints, but also as a shrewd maneuver to further strengthen ties with Cold War allies. ESA agreed to construct a permanent scientific laboratory, somewhat like Spacelab. Canada created a robotic arm for servicing the station's exterior. Japan built a scientific laboratory that included a small robotic arm and an area exposed to space for experimentation. While JSC's Mission Control Center served as the primary control room for space station operations, each space agency agreed to build its own control center to oversee its contribution. Japan, for instance, built a Space Station Operations Facility at the Tsukuba Space Center in the Ibaraki prefecture, completed in 1996.[71]

On 11 February 1988, the NASA Space Station Working Group and the ESA Council drafted an MOU regarding a proposed space station and enumerating each space agency's contribution to the space station. The low-earth orbit station remained flexible for a number of projects, including science, earth observation, storage, and service as a staging base for future space efforts. The use of the station and its elements was equitable for each organization involved. They also debated issues

International Space Station, 2011. Courtesy of NASA, http://spaceflight.nasa.gov/
gallery/images/station/assembly/html/s133e010447.html.

on operations, safety, the crew, and communications, among other
things.[72]

After a series of redesigns, and encumbered with an ever-decreasing
budget, NASA finally cancelled the Freedom project in 1993.[73] NASA
and the other space agencies instead continued in their efforts to build
an even larger space station, this time with the help of the post–Cold
War Russian space agency. Russia, in fact, was a chief partner, contributing five separate modules. Aside from those additions, approximately
75 percent of the previous space station concept remained the same.
The orbit also was changed to accommodate the Russian launch area
more easily. The partners eventually renamed the new project: the International Space Station (ISS).

Russia's inclusion in the ISS did spark some controversy, beyond the
implications of former enemies working so closely together. ESA, Canada, and Japan each expressed dismay with NASA for not consulting
them before making such a drastic decision. Some officials remarked
that it showed American arrogance. Others felt that their partnership

status had been disrespected. Perhaps it was a result of the easing of international tensions after the Cold War, and a potentially diminished need for good relations with the other nations. NASA eventually smoothed over any problems, and the various space agencies completed their contributions to the ISS.

Russia launched the first component, *Zarya* ("Dawn"), into orbit on 20 November 1998. This central piece provided power, control, communications, and docking capabilities for the early construction phases. NASA's *Unity* module, a connecting node akin to a hallway, joined *Zarya* on 6 December 1998. Financial problems for the Russian space agency continually created problems for the ISS, pushing back schedules and compelling others, especially NASA, to provide monetary help.

After a delay of almost a year, Russia finally launched *Zvezda* ("Star") on 12 July 2000. This critical module included life support, navigation, propulsion, and living quarters. It allowed the first crew to begin living onboard the ISS on 31 October 2000. Since that date, humanity has continuously had a presence in space.

Construction of the ISS remained relatively steady for the next two years. NASA launched the scientific laboratory *Destiny* on 7 February 2001. Canada attached its robotic arm in April 2001. NASA's airlock *Quest*, added in July 2001, allowed the astronauts to conduct extensive EVAs. Russia's *Pirs* ("Pier"), which includes more docking ports and an airlock for Russian cosmonaut suits, was launched on 14 September 2001. The ISS also grew with various additions to the truss system and solar arrays. Additions came to a halt, however, after the loss of the shuttle *Columbia* on 1 February 2003.

When NASA placed the shuttle fleet on hold to determine the cause of the *Columbia* disaster, the assembly of the ISS had to be slowed too. For more than two years, only Russian rockets could replace crews and conduct resupply missions. Full work on the station began again with the addition of external stowage platform (ESP)-2 on 26 July 2005. Shortly thereafter, NASA announced the planned retirement of the space shuttle fleet in 2011, setting a formal deadline for completion of the ISS.

ESA's laboratory, *Columbus*, finally joined the station in February 2008. Japan's laboratory, *Kibo* ("Hope"), required three different launches for its components in March and May 2008 and July 2009. Russia added one small research module, *Poisk* ("Search"), in November 2009, and another, *Rassvet* ("Dawn"), in May 2010. NASA also contributed two more nodes, *Harmony* in October 2007 and *Tranquility* in February 2010, as well as completing the truss structure and solar arrays. With the STS-134 mission, the international team completed assembly of the ISS in May 2011.

One other European contribution requires mentioning. The Italian Space Agency built three multipurpose logistics modules (MPLM): *Leonardo, Raffaello,* and *Donatello.* Designed to fit inside the cargo bay of the shuttle, MPLMs served as large shipping containers for cargo to the ISS. Between March 2001 and July 2011, these MPLMs flew a total of twelve times. In early 2011, NASA reconfigured *Leonardo* as a permanent multipurpose module, providing storage for supplies and waste for the station. *Leonardo* joined the ISS permanently on 1 March 2011.

Each partner was required to build a control center to monitor its components. Japan has its control center in Tsukuba, around 70 kilometers (45 miles) northeast of Tokyo. Canada monitors the robotic arm from the Mobile Servicing System Operations Complex (MOC) in Saint-Hubert, Quebec. ESA built a Columbus Control Centre in Oberpfaffenhofen, Germany. Because Russia built two of the first three components, primary control began at the Russian control center near Moscow. ISS control currently resides in JSC's Mission Control Center.[74]

* * *

Space has a complex history of international cooperation and competition, particularly when space agencies are used as extensions of their nation's foreign policy. Adversaries have become partners, and vice versa. Sometimes space agencies work together on one project while competing in another area. The control centers and the flight controllers have played a vital role in this story. Anytime space agencies collaborate on a mission, be it Giotto, ASTP, or ISS, the controllers must also work together to monitor the spacecraft. While space agencies will continue to pursue their own projects, as long as budgets remain shrunken, cooperation will be increasingly the norm. The International Space Station stands as the prime example of the necessity for countries to work together as they reach beyond earth.

5

TRACKING NETWORKS

During John Glenn's Mercury flight, *Friendship 7*, the controllers in Florida noticed a potential problem. A warning light indicated that the heat shield and landing bag for the capsule were out of position. If the heat shield was damaged in any way, heat from the atmosphere during reentry could seep into the capsule and destroy it. As a precaution, Mercury Control requested that each of the remote sites monitoring the spacecraft across the world pay particular attention to the warning light. When the first remote-site capcom asked Glenn about the landing bag switch, he paid little heed. But as each subsequent remote site did the same, he realized that something must be amiss, even if he noticed no warnings or lights in the capsule itself.

Finally, as a safety measure, Mercury Control requested that Glenn keep the retrofire rocket pack, which was positioned over the heat shield, locked on longer than usual. In a typical flight, it was jettisoned immediately prior to reentry. If the pack stayed on during reentry, it would burn up in the atmosphere, and pieces flying off it had the potential to fatally damage the capsule. The controllers argued that it was safer to keep it on in case the straps were the only things holding the heat shield in place.

During reentry, Glenn did notice pieces of something flying past the window, fearing they might be from the heat shield itself. Fortunately, the warning light turned out to be a simple glitch, and the retrofire pack did not damage the heat shield. Glenn returned home safely.

The *Friendship 7* flight provides an excellent example of the necessity for remote-site antennas around the world. Because of the curvature of the earth, it is impossible for one antenna to maintain contact with an orbiting object, unless it is in geosynchronous orbit. Spaceflight agencies must build antennas around the world in order to maintain constant communication with their spacecraft. Consequently, communications networks serve as vital links between the ground segment and the spacecraft.

Communications networks also serve as integral aspects of international diplomacy for space agencies. The space programs have to find appropriate areas for the equipment in friendly countries. They also must coordinate the construction of and work on the antennas in the various locations, which are typically staffed by locals. The foreign relations aspects played an especially important role during the Cold War, when space programs were an avenue for superpowers to showcase their prominence.

Each control room developed its own communications network. JSC has primarily used the Manned Space Flight Network (MSFN), which became the Space Tracking and Data Network (STDN) and transitioned into the Tracking and Data Relay Satellite System (TDRSS). JPL operates the Deep Space Network (DSN). ESOC maintains the European Tracking Network (ESTRACK). Although those are their primary communications networks, each control center has utilized the others' networks at some point to monitor their missions. In general, the networks of JPL and ESOC are fairly similar in makeup and use.

Each of the networks is largely concerned with tracking spacecraft and handling communications between controllers on the ground and objects in space. The only major difference arises in the scope of their work. The STDN and TDRSS track near-earth spacecraft, especially the manned NASA programs. The DSN, as the name suggests, monitors deep space missions for JPL and other space organizations. ESTRACK, throughout its history, has served in both capacities.

The accumulation of data differs greatly between near-earth and deep space objects. For deep space missions, the network generally has prolonged amounts of time of direct contact with the satellites, because of the distance, and prolonged times of no contact as well. Because signals must travel over vast distances, their strength decreases while there is the commensurate time difference between signal start and acquisition. Near-earth satellites, on the other hand, have much stronger signals, but the passes over antennas are much shorter. The network usually needs more stations because the decreased line of sight, in comparison to deep space missions, means that each antenna provides less coverage of the sky. Both types of missions have their own communications advantages and disadvantages that, in turn, inform the placement of the network sites.[1]

MANNED SPACE FLIGHT NETWORK TO TRACKING AND DATA RELAY SATELLITE SYSTEM

Mission control cannot speak with the astronauts directly, so some kind of intermediary is necessary for the transmission of communications. In order to make the links as effective as possible, NASA and other space agencies adhered to a number of considerations when setting up their networks. One of the most important was spacing the antennas in appropriate areas for as much coverage as possible. Because of the lack of land in certain areas of the earth, this proved difficult, but they did their best to make up for any gaps, including using ships. Maintaining stations in foreign nations could lead to interesting international political intrigue, which meant that NASA preferred to deal with countries friendly with the United States.

NASA's manned spaceflight program used three different networks of communications for various missions. At first, the Space Tracking and Data Acquisition Network (STADAN) was based out of Goddard Space Flight Center (GSFC) in Greenbelt, Maryland. It was primarily concerned with near-earth orbit satellites. The Deep Space Network (DSN) is still controlled through the Jet Propulsion Laboratory. The Manned Space Flight Network (MSFN) served as the primary communications network for JSC during the Mercury, Gemini, Apollo, and Skylab missions. Following Skylab, the STADAN and MSFN merged to form the Spaceflight Tracking and Data Network (STDN). In 2009, NASA formally replaced this with the Tracking and Data Relay Satellite System (TDRSS). The NASA Communication Network (NASCOM) served as a relay network between the antennas and the control centers for each of the networks.

STADAN began as the Minitrack Network in 1957, the first American satellite-tracking network. The original Minitrack Network included six South American sites: Havana, Cuba; Quito, Ecuador; Lima, Peru; Antofagasta, Chile; and Santiago, Chile. Sites in North America included the following: Blossom Point, Maryland; San Diego, California; Fort Stewart, Georgia; Coolidge Field, Antigua; and Grand Turk Island. NASA built another site in Woomera, Australia, shortly after the network came online. Over the course of its history, STADAN site locations have been fairly fluid, depending on the needs of the satellites.[2]

Minitrack was operational in time for the first *Sputnik* launch. After some calibrations, the different sites were able to track *Sputnik*, giving them some practice and experience before NASA's first satellite flights. After its foundation, NASA integrated Minitrack immediately into the space agency, along with what became the Goddard Space Flight Center (GSFC). Between 1958 and 1962, Minitrack made a number of changes regarding station locations. NASA added sites in Fairbanks, Alaska; East Grand Forks, Minnesota; St. John's, Newfoundland, Canada; and Winkfield, England. They also closed down the site in Antigua. Following the Cuban Revolution in 1959, NASA moved the Havana site to Fort Myers, Florida.[3] NASA could not accept the risk of having a remote site located in a communist country sympathetic to their Cold War rivals.

Between 1962 and 1966, Minitrack slowly transformed into STADAN with newer, bigger antennas and consolidated some of its sites. Fairbanks received the first twenty-six-meter (eighty-five-foot) antenna in 1962. As larger antennas provided better coverage and stronger signals, NASA required fewer sites. By 1965, STADAN had twenty-two sites around the world, though NASA closed six the following year. Just seven years later, NASA consolidated STADAN to ten major sites: Canberra, Australia; Fairbanks, Alaska; Goldstone, California; Quito, Ecuador; Santiago, Chile; Rosman, North Carolina; Fort Myers, Florida; Winkfield, England; Johannesburg, South Africa; and Tananarive, Madagascar.[4]

The Manned Space Flight Network had origins in pre-NASA work but actually began after the creation of the space agency and the beginning of Project Mercury. NASA realized that in order for its missions to have proper ground control, it needed both a network of communications antennas and a way of transferring those communications back to mission control. NASA thus created what became the MSFN and NASCOM. On 30 July 1959, NASA contracted with Western Electric, Bell Laboratories, Bendix, Burns and Roe, and IBM to build the first series of radars, the Mercury Network. NASA also decided to construct a computer complex, consisting of a primary IBM 7090 computer and one backup, at Goddard to handle incoming communications and link to the Mercury Control Center. NASA built a backup computer center in Bermuda.[5]

The early network included sites in Cape Canaveral, Florida; Grand Bahama Island; Grand Turk Island; Bermuda; Grand Canary Island; Kano, Nigeria; Zanzibar (Tanzania); Muchea and Woomera, Australia; Canton Island; Kauai, Hawaii; Port Arguello, California; Guaymas, Mexico; White Sands, New Mexico; Corpus Christi, Texas; and Eglin Air Force Base, Florida. They also maintained two ships with tracking equipment: *Coastal Sentry Quebec* in the Atlantic Ocean and *Rose Knot Victor* in the Indian Ocean. NASA completed construction on the sites between April 1960 and March 1961.[6] Flight controllers, engineers, and astronauts staffed each of the sites, giving them valuable experience. The network worked well throughout the Mercury program.

A typical site consisted of four controllers. The capsule communicator served as both the capcom and the flight director for their site. He communicated directly with both the astronaut in space and Mercury Control. The capcom, in the most critical sites, was an astronaut, thereby setting the precedent continued at JSC. A maintenance operations supervisor oversaw the equipment at the site. A systems monitor examined the spacecraft's systems. Finally, an aeromedical monitor, staffed by a flight surgeon, processed the astronauts' medical information.[7]

Before Gemini, NASA realized the network needed more computing capabilities for future missions and installed UNIVAC 1218 computers at each of the sites. These on-site computers allowed the controllers in mission control to receive more information faster, which greatly aided their work. This, along with the enhanced mission control in Houston, allowed NASA to keep most of its essential controllers and engineers centralized rather than "in the field" during missions. As a result, most of the senior controllers remained in Houston while newer members manned the remote sites to gain experience. NASA also added sites in Antigua; Ascension Island; Pretoria, South Africa; Tananarive, Madagascar (from Zanzibar, after a revolution in 1964); Wallops Island, Virginia; and Goddard.[8] NASA stationed the two tracking ships in the Pacific Ocean for greater coverage.[9]

As JSC transitioned to Apollo, NASA recognized an even greater demand on the network. NASA could neither transmit signals to nor receive them from the moon by the near-earth focusing network, at least not reliably. Also, when the Lunar Module (LM) separated from the Command/Service Module (CSM), the network did not have the ability to track both spacecraft accurately. The network needed a massive upgrade or NASA needed to pair it with another network or both. After significant consideration, NASA decided to pursue both measures. The MSFN received upgrades, but it also had support from the STADAN and DSN.

Because NASA already planned for DSN to track lunar missions, like Surveyor, the infrastructure was largely in place to aid with Apollo. NASA built twenty-six-meter (eighty-five-foot) antennas at the most critical MSFN sites in order to accommodate lunar communications. Many of these were near DSN sites so that they could more easily complement each other.[10]

One major change in the structure of the network came with the addition of radar ships. NASA stationed one ship in each of the Pacific, Atlantic, and Indian Oceans, while two served during reentry in the Pacific. The Apollo Network also utilized a series of airplanes called Apollo Range Instrumentation Aircraft (ARIA), which served as highly mobile communications relays. The mobility of ships and aircraft was especially important for human spaceflight when constant communication was more crucial than for robotic spacecraft.[11] Flight controllers from Houston did not operate the various sites during Apollo. Instead each site had its own staff, usually contractors, because NASA transferred the information to mission control almost simultaneously.[12]

In the mid-1970s, following Skylab, NASA combined the STADAN and the MSFN to create a unified STDN. While this reduced costs, NASA quickly realized that the STDN remained too manpower-intensive. Each remote site required operations, maintenance, and logistics personnel. Overseas locations were especially expensive. As a result, NASA sought a more cost-effective method of communicating with the astronauts in space. Shortly after the creation of the STDN, NASA began to plan for a new Tracking and Data Relay Satellite System (TDRSS) as a replacement. In essence, two geosynchronous satellites could cover communications for nearly the entire orbit of a shuttle or other near-earth spacecraft. The network also included a single ground site at Goddard to consolidate the system.

With this new network, the remote sites became unnecessary, thereby reducing the long-term expense. TDRSS did require a large initial cost in order to pay for the new satellites and equipment. To save money, NASA contracted out the satellites. Thus, Western Union Space Communications Inc. won the initial contract in December 1976. Western Union subcontracted the production of the satellites to TRW Inc.[13]

Subsequently, NASA had been slowly closing down overseas tracking stations, until the TDRSS came online. The first shuttle flights could not be tracked for their full flights. In fact, JSC could communicate with the astronauts during only 40 percent of the missions. Fortunately, no major problems arose that could have led to disaster without full coverage.[14]

NASA placed the first Tracking and Data Relay Satellite (TDRS) into orbit in 1983 during STS-6.[15] NASA then placed five more first-generation satellites, built by TRW, into orbit using the shuttles in the 1980s and 1990s. NASA launched three second-generation satellites, built by Boeing, on Atlas rockets between 2000 and 2002. TDRSS interacts with three ground-based locations, with a new primary station in White Sands, New Mexico, and others in Guam and at Goddard. The entire system currently has eight satellites in orbit, although only three are designated as primary, and are thus continuously in use. NASA placed them in a constellation in geosynchronous orbit to provide the maximum amount of coverage.[16]

The primary satellites maintain orbit on the equator at 41 degrees west and 171 degrees west. Their geosynchronous orbits allow them to maintain 85 percent coverage for spacecraft below 1,200 kilometers (745.6 miles) altitude. Spacecraft in orbits up to 12,000 kilometers (7,456.5 miles) altitude gain complete coverage. For reference, the shuttle usually orbited at altitudes between 304 kilometers and 528 kilometers (190 to 330 miles) and the International Space Station maintains its orbit between 370 and 460 kilometers (230 to 286 miles).[17] The coverage creates a small zone of exclusion (ZOE) where the satellites cannot provide communications coverage for the shuttles or ISS. The additional TDRS in orbit can be powered up to provide an emergency bridge over the ZOE.

Each TDRS has ten years of attitude control fuel, so they need to be replaced periodically.[18] NASA plans to launch two third-generation satellites.[19] *TDRS-K* successfully launched on 30 January 2013 on an Atlas V rocket, while *TDRS-L* was launched on an Atlas V rocket on 23 January 2014.[20] The TDRSS should serve human spaceflight well for at least the next decade.

DEEP SPACE NETWORK

The origins of JPL's Deep Space Network are contemporaneous with the launch of *Sputnik* by the Soviet space program. William Pickering of JPL recognized that there was too much electrical interference in the Pasadena area to track *Sputnik* accurately, so they had to use the San Gabriel Valley Radio Club in Temple City, approximately fourteen miles southeast of JPL. He stated that JPL needed a site in the desert with a tracking station, which led to the large antenna eventually built in Goldstone.[21]

Because the DSN communicates with spacecraft much farther away than earth orbit, it must have a different makeup from that of the manned spaceflight network. Signals are weaker, and exact locations and trajectories are more difficult to pin down. In fact, signal strength changes inversely to the square of the distance. In other words, communications from Neptune are one hundred times weaker than from Mars, or ten billion times weaker than with a geostationary satellite. To ensure continued receptivity of radio signals, the DSN must constantly update its equipment, hardware, and software.[22]

In general, the distance of the spacecraft being tracked and the coverage of the antennas allow JPL to utilize fewer antennas than JSC. The DSN thus can be more selective in site placement. Regardless, JPL has benefited from conveniently placed allies of the United States.

Radio astronomy, the basis of tracking and data acquisition for spacecraft, can be traced to Karl Jansky. In 1931, this electrical engineer detected extrasolar radio signals by using a radio receiver consisting of an antenna array rotated by four Model T wheels and a motor. From such humble beginnings, tracking and detection improved dramatically over the next two decades.

By November 1958, JPL had connected an early network of antennas, called Microlock, to the first control room at the center. Early Microlock stations were connected to the control area through telephones and teletype. Microlock was used mainly for near-earth objects, especially missiles. JPL also established an early deep space network of antennas called Tracking and Communication Extraterrestrial (TRACE). Microlock was especially important for establishing the first series of antennas for tracking. For instance, JPL constructed four stations in 1958 for Explorer: Cape Canaveral, Florida, Earthquake Valley, California, the University of Malaysia, Singapore, and University College in Idaban, Nigeria.[23] JPL's connection to Caltech must have played a role in the location of the two foreign sites.

The early TRACE network, on the other hand, did not extend worldwide and therefore offered only limited coverage. JPL built the first three stations in Goldstone, California, Cape Canaveral, and Mayaguez, Puerto Rico. The Goldstone site by this time did include a twenty-six-meter (eighty-five-foot) antenna, one of the largest at the time. JPL chose Goldstone for the center of its Deep Space Network for a few reasons. First, it was relatively close to Pasadena. Second, its geography, essentially a bowl surrounded by hills, keeps it mostly radio silent. Perhaps most fortuitously, the United States Army owned the land, and JPL had maintained a close relationship with the Army since its early days. Goldstone was ready for use by the end of 1958.[24] JPL named this first antenna Deep Space Station 11, or DSS 11. DSS 11 first tracked a satellite in March 1959, as *Pioneer 4* became the first American spacecraft to reach the moon.[25]

In 1958, JPL also proposed a more formal network of three stations to track deep space objects. Two stations were constructed, one in Nigeria and one in the Philippines, to work along with the station in Goldstone. Although originally accepted, later that year a Department of Defense official, Dr. Donald Quarles, questioned the two overseas locations. After further studies, JPL amended the proposal to change the locations to Spain or Portugal and Australia, both of which increased coverage for deep space missions.

NASA built a new twenty-six-meter (eighty-five-foot) antenna at Goldstone in 1960 to aid in transmitting capabilities for the DSN.[26] NASA completed construction of the second deep space antenna at Island Lagoon, Australia, in September 1960. A third antenna, planned for Spain, was moved to South Africa. This site, about forty miles north of Johannesburg, was completed in July 1961. With that antenna, JPL had nearly worldwide coverage, and the Deep Space Instrumentation Facility (DSIF) was born.[27] Like those of STDN, these remote sites were largely manned by contractors. Collins Radio Company, for instance, provided employees for Johannesburg.[28] Two years later, on 24 December 1963, JPL paired the DSIF with ground communications and the early SFOF to become the Deep Space Network (DSN), under the direction of Eberhardt Rechtin.[29]

NASA realized that it was overworking the DSN in the mid-1960s. Meetings to schedule time on the network were often quite contentious, with priority regularly going to those with the loudest voices rather than those with the most pressing needs. NASA set out to build a second network of stations to be paired with the DSN.[30] As a result, NASA built another antenna for the DSN in Australia. Located in Tidbinbilla Valley, near Canberra, this antenna was complete and online by March 1965. NASA chose this site because it was relatively noise-free and there had previously been federal land set aside in the area, which made negotiations easier. Another station was built in Robledo de Chavela, Spain, near Madrid. This station was completed in July 1965. Yet another station came online in June 1966. Located on Ascension

Island in the South Atlantic Ocean, it allowed for more coverage between the Americas and Africa.[31]

These three stations provided more coverage for the network and played an integral role in the DSN's backup coverage for the impending Apollo program. NASA decided to use the DSN as support for Apollo and wanted to have as many redundancies as possible to avoid potential loss of life.[32] The DSN is also the primary communications network for projects operating more than sixteen thousand kilometers (ten thousand miles) from the earth. Because the moon is over 322,000 kilometers (two hundred thousand miles) away from the earth, NASA naturally called upon the DSN for the Apollo missions, as well as other missions to the moon, including Surveyor.[33] JPL further updated the Goldstone site by constructing a new sixty-four-meter (210-foot) antenna for increased deep space communications. This large antenna, named DSS 14, was finished on 16 March 1966, slightly behind schedule.[34] The new antenna proved crucial for the deep space missions in future years, beginning with the first missions to Mars.

By 1964, the DSN facilities had teletype capabilities. In fact, teletype data was directly input into the computers at the facilities, making punch cards unnecessary. Also in the mid-1960s, before the Surveyor missions to the moon, they installed a microwave system to aid communications between the network in Goldstone and JPL itself. This system compensated for the increased amount of data anticipated for Surveyor and other future projects.[35]

With the Pioneer missions in the 1960s, the DSN began supporting missions with control centers off JPL property. Ames Research Center housed the control room, and JPL provided the data, an aspect of sharing information rarely part of the STDN. Pioneer also created some problems by maintaining data transmission for years after its expected termination. JPL received the last signal from *Pioneer 10* on 23 January 2003. The DSN thus had to remain vigilant much longer than expected.[36] These experiences provided important data for JPL's future missions. In this time frame of the 1960s, the DSN changed from supporting single missions to supporting multiple missions.

During the late 1960s, the DSN became almost entirely absorbed with lunar-exploration missions. This began with Surveyor and Lunar

Orbiter, but it continued with the backup support of the Apollo missions. While the DSN supported multiple missions, it was not until later that it could do so simultaneously. Other programs within NASA were clearly secondary to the missions to the moon. During the near-earth portions of the missions, the DSN did receive help from the Manned Space Flight Network in collecting data.[37]

As the DSN supported the Apollo missions, it began in the 1970s to transition toward having more multimission compatibility. JPL added wings to each of the DSN station buildings to help support the Apollo missions. It also prepared for more ambitious missions to the outer planets.[38] In order to deal with the influx of communications created by multiple missions, the DSN added a Multimission Telemetry System in 1969.[39] During this period, the DSN also consolidated its sites. The Ascension Island stations, for instance, were transferred to Goddard Space Flight Center control in November 1969. The DSN also dismantled the Woomera, Australia, station and took over the MSFN station at Honeysuckle Creek, near Canberra, in 1973. The following year, the DSN closed the station near Johannesburg, South Africa. The DSN also constructed two new sixty-four-meter (210-foot) antennas to supplement the one at Goldstone. The first, operational in April 1973, was constructed in Tidbinbilla, Australia, also near Canberra. The second, operational in September 1973, was constructed in Robledo, Spain, near Madrid. These new antennas were named DSS 43 and DSS 63, respectively.[40]

Even with the new larger antennas, the DSN still utilized the twenty-six-meter (eighty-five-foot) antennas for a variety of reasons. For instance, after launch and during the first phases of a mission, a spacecraft's angular movement cannot be tracked by the large antennas. They relied on the smaller dishes during the first phases of missions. After the spacecraft had been acquired and established, the DSN could switch to the sixty-four-meter (210-foot) antennas for the remainder of the missions.[41]

The Surveyor missions of the mid-1960s were the first ones to be dependent solely on the DSN for communications.[42] The Mariner Mars 1969 mission conducted flybys of the planet for various observations in preparation for future missions to the planet. The SFOF was able to utilize the new sixty-four-meter (210-foot) antennas in order to obtain information more quickly from this mission, as well as real-time television pictures and other data.[43] In some ways, Mariner Mars 1969 proved the importance of the larger antenna and the possibility of future scientific knowledge from Mars.

To demonstrate how missions from other centers relied on the DSN for communications, between 1961 and 1974, the DSN supported the Ranger, Mariner, Pioneer, Apollo, Surveyor, and Lunar Orbiter programs. Only Ranger, Mariner, and Surveyor were managed by JPL. Ames Research Center controlled Pioneer; Langley Research Center directed Lunar Orbiter; and JSC controlled Apollo.[44] In later years, the DSN provided communications support for projects from numerous foreign space agencies, including those of Japan, Russia, India, and Europe, thereby furthering the role of JPL in NASA's foreign relations initiatives.

The Viking missions of the mid-1970s presented new issues for the DSN to handle. Two spacecraft were launched nearly simultaneously. Each spacecraft then separated into an orbiter and lander as it neared Mars, meaning that at times the DSN had to track four spacecraft in close proximity. The new sixty-four-meter (210-foot) antennas played an integral role in the success of Viking.[45] These antennas were also largely responsible for communications with *Helios 1* for twelve years, from launch in 1974 to loss of signal due to the termination of operations by receivers in 1986.[46]

Between 1978 and 1980, three of the twenty-six-meter (eighty-five-foot) antennas were upgraded to thirty-four-meter (112-foot) antennas. These new antennas allowed for greater range and more coverage of different frequencies for deep space missions. The DSN added thirty-four-meter (112-foot) high-efficiency antennas to Goldstone and Canberra in 1984, and to Madrid in 1987. These new antennas, working with those already in place, widened the frequencies and created more flexibility in the network.[47]

In 1981, DSS 11 in Goldstone was decommissioned after supporting missions for twenty-three years, including Pioneer, Mariner, Lunar

Surveyor, and Voyager, among others. The site was named a national historic landmark on 27 December 1985, in recognition of its central contribution to so many lunar and planetary missions.[48]

In the mid- to late 1980s, the DSN aided Australia and the European Space Agency to upgrade the equipment in Canberra. Because all three space organizations anticipated using these facilities in the near future, each contributed something so that Canberra could better serve a variety of space projects. The DSN similarly upgraded the stations in Spain and Goldstone simultaneously.[49]

While the sixty-four-meter (210-foot) antennas had performed well for a few decades, the DSN called for upgrades to its largest equipment. After some deliberation, especially regarding cost, the administration agreed to replace those three antennas with even larger seventy-meter (230-foot) antennas. It took almost five years to complete the new antennas, but by May 1988, the three DSN stations were finished.[50]

By the mid-1980s, the DSN's reputation had grown strong enough that it became a source of important international cooperation for JPL and NASA. Usually this cooperation came about informally, until a directive in 1991 formalized how the DSN and other communications networks conducted arrangements for the use of networks by foreign agencies.[51]

During the 1980s, NASA also worked on a reconfiguration of its communications networks. As a result, elements of the Ground Spaceflight Tracking and Data Network, operated at the Goddard Space Flight Center, were integrated into the DSN for greater coverage. This consolidation also occurred for monetary reasons, as NASA continued its efforts to deal with a decreasing budget.[52]

In the late 1990s, the DSN built new thirty-four-meter (122-foot) antennas at each of its sites in order to replace the oldest antennas.[53] The DSN is now sensitive enough to detect natural emissions of electromagnetic radiation, including stars, gas clouds, and even Jupiter. Naturally, JPL has sponsored analysis of DSN readings from these natural discharges.[54]

The DSN continues to serve as the most prominent antenna network for spaceflight. Its operations dominate the Operations Room of JPL. While the DSN remains the most famous communications network,

Europe's own network has an impressive history for both deep-space and near-earth communications.

ESTRACK

The importance of ESTRACK to the success of ESOC cannot be overstated. Indeed, one author has compared ESOC without ESTRACK to a carriage without a horse, or, perhaps more appropriately, to a car without an engine.[55] A more contemporary comparison might be a cell phone without a tower.

ESTRACK has proven to be one of the most consistently variable aspects of ESA. Not only has the number of stations changed rapidly, but their locations have necessarily changed in order to suit mission needs. Early programs were focused on polar observations, thus the network focused on stations in higher latitudes. As the focus changed to earth observations and deep space missions, new stations in more central locations have replaced the higher-altitude stations.

With early satellites remaining in high inclinations, the first ground stations needed to be located in high latitudes to maintain maximum connection. Thus, ESA selected Redu, Belgium; Spitzbergen, Norway; Fairbanks, Alaska; and the Falkland Islands as the locations for its first four ground stations. The Redu station also obliged the need for a station located near the control center at the European Space Research and Technology Centre (ESTEC) in Noordwijk, the Netherlands.[56] Like the other networks, ESTRACK locations proved important political tools.[57] Early in the development, ESA included a two-way teleprinter to connect the ground stations with the control center for immediate communications.[58] With the lack of high-latitude programs beginning in the mid-1970s, ESA decided to close the Falkland Islands station on 31 December 1973 and the Spitzbergen station on 16 April 1974.[59] Further changes led to the Fairbanks facility officially closing in August 1977, which left only Redu of the original four stations. New ground stations, however, had been built in Kourou, French Guiana; Villafranca, Spain; Odenwald, Germany; Fucino, Italy; and Malindi, Kenya.[60] The Villafranca station, online on 12 May 1978, also served as a dedicated control center.[61] The Fucino station has been used as part of ESTRACK sporadically, depending on the needs of the various programs, throughout its existence.[62] ESA built a new station in Carnarvon, Australia, in June 1980.[63] Another station was added to the network in Ibaraki, Japan, in 1984.[64] Between 1987 and 1988, the Carnarvon station moved to Perth, Australia.[65] By 1988, ESTRACK had also added stations in Kiruna, Sweden, and Maspalomas, Canary Islands.[66]

Ground stations in 2004 included those in Kiruna, Redu, Villafranca, Cebreros (Spain), Maspalomas, Kourou, and Perth. ESOC also had access to cooperative tracking and launch stations in Svalbard, Norway; Plesetsk, Russia; Baikonur, Russia; Tsukuba, Japan; Malindi; Santiago, Chile; Goldstone; and Canberra.[67] As of April 2007, ESTRACK included those same stations, with the addition of New Norcia in Australia. The New Norica and Cebreros stations included thirty-five-meter (115-foot) antennas as part of the newly formed European Deep Space Network.[68] ESTRACK added a new Santa Maria Tracking Station in the Azores (Portugal) in 2008.[69]

Like JPL's DSN, ESTRACK has been an important conduit for relations with other space agencies. ESA has provided tracking for numerous spacecraft from various other space programs, including NASA, Russia, Japan, India, and China. In return, ESA has been allowed either to establish stations in those countries, such as Japan and the United States, or to use their own networks. Communications with the Japanese Ibarak station could be especially troublesome, however, because few of their controllers spoke English.[70] Perhaps most importantly, ESA has utilized JPL's DSN for numerous missions. In many ways, this cooperation and codependence has fueled further relations between ESA and NASA. The CCSDS Space Link Extension (SLE) serves as a standard interface system between various ground stations and control centers across the world, particularly between ESTRACK and NASA's DSN.[71]

Tracking varies widely depending on numerous variables for each mission. As an example, a typical high-latitude orbit may be visible for ten minutes during ten out of every fourteen orbits. In stark contrast, deep space missions may have contact for up to ten hours at a time, though blackout periods may last days or even weeks. The controllers must anticipate these changes in frequency of contact and adapt in

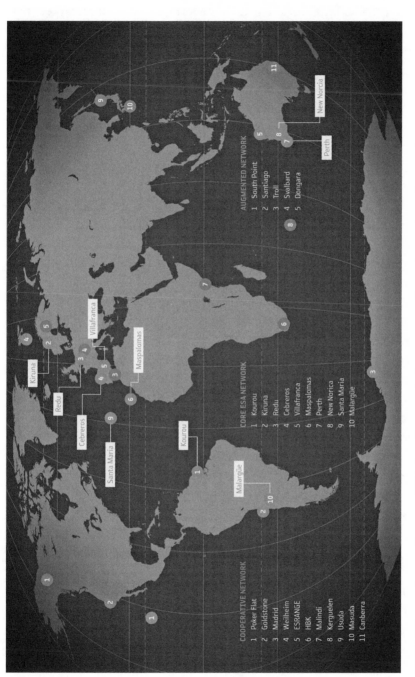

COOPERATIVE NETWORK

1 Poker Flat
2 Goldstone
3 Madrid
4 Weilheim
5 ESRANGE
6 HBK
7 Malindi
8 Kerguelen
9 Usuda
10 Masuda
11 Canberra

CORE ESA NETWORK

1 Kourou
2 Kiruna
3 Redu
4 Cebreros
5 Villafranca
6 Maspalomas
7 Perth
8 New Norcia
9 Santa Maria
10 Malargüe

AUGMENTED NETWORK

1 South Point
2 Santiago
3 Troll
4 Svalbard
5 Dongara

ESTRACK network, 2013. Courtesy of ESA, http://www.esa.int/spaceinimages/Images/2013/01/Estrack_network.

order to maximize data return.[72] One former controller remarked that timing was critical for deep space missions.[73]

Since ESOC controls spacecraft everywhere from near-earth to deep space, ESTRACK must remain flexible to track the wide variety of altitudes. In many ways, ESTRACK can be considered a mix of the STDN and the DSN. It must utilize larger antennas like DSN for deep space missions, but it also must maintain more sites like STDN for coverage of near-earth satellites.

A typical ESTRACK ground station includes a Main Equipment Room (MER), which contains the necessary hardware for telemetry, telecommands, and data processing, among other essential operations. An Antenna Equipment Room (AER) houses the antenna servo-system, air conditioning, and other ancillary systems, such as lighting.[74] The structure and configuration of all ground stations are intentionally basic for ease of transition and upgrade.

Both ESTRACK and ESOC have gained an ever more positive reputation over the decades. ESTRACK is especially renowned for its now remotely controlled stations. While DSN continues to operate its antennas with on-site personnel, ESTRACK has upgraded its facilities to run automatically by computers. Automation, it is argued, limits the possibilities for human error. It also reduces costs by reducing needed personnel. A centralized ESTRACK Control Centre (ECC) in the Operations Control Centre oversees all ESTRACK operations. If a problem occurs that cannot be fixed from the control room, on-call technicians can reach the ground station within an hour. The fully automated remote sites have proven so effective that JPL has begun to study them as possible upgrades to their DSN ground stations.

DIPLOMACY

The selection of antenna sites around the world has played a role in NASA's foreign diplomacy, particularly during the Cold War. If multiple locations were in the running for expansion, often NASA made its selection based on the foreign policy needs of the United States. NASA might choose a friendly country over a neutral country, or use the opportunity as a bargaining chip to cultivate better relations with a neutral country.

For instance, as previously mentioned, JPL proposed to expand its early TRACE network into Nigeria and the Philippines. While NASA approved of this arrangement, the Department of Defense stepped in and argued that other sites might be more beneficial. A second study changed the locations to the Iberian Peninsula and Australia. Thus, the DOD influenced NASA to change the locations of the new sites from poorer, relatively neutral countries to richer, stronger allied nations.

Sometimes, remote sites also had to dodge real-life events. Originally the STADAN included a site in Havana, Cuba. The Cuban Revolution in 1958 and 1959 led to a new government, with Fidel Castro in power. The Revolution and Castro's anti-American political beliefs forced NASA to build a replacement in Fort Myers, Florida. Similarly, the country of Zanzibar erupted into revolution in 1964. Not only did the new government shed Arab dominance and merge with Tanganyika to form Tanzania, but the revolution also necessitated a replacement remote site in Tananarive (now Antananarivo), Madagascar. It should be noted that fears of a communist takeover also influenced this move.

Remote sites could also be used for public relations. Glenn's *Friendship 7* also provides an example of this. As he flew over Australia, the city of Perth turned on as many lights as possible, as did a British Petroleum site south of the city. When the remote-site capcom, L. Gordon Cooper, asked Glenn if he could see anything, he remarked that the lights could be clearly seen and he thanked the citizens of Perth. This served as an example of NASA reaching out to civilians in other countries and the interconnectedness of space travel. Perth even became known as the "City of Light" because of this moment.[75]

Unfortunately, visiting foreign countries does not always bring out the best in people. NASA officials sometimes made statements that can be taken as either veiled or implied racism. For instance, after visiting Nigeria and Zanzibar, one wrote about the "native population which is capable of believing almost anything, and getting quite excited about it." The industrialized American disparages the naïve native who can be easily tamed by advanced technology. This is eerily reminiscent of eighteenth-century explorers looking down on Pacific Islanders who could be bought by the "simplest" technology, like an iron nail or a button, or who were awed by skyrockets. This may not have been the intent of the official at the time, but the modern reader can certainly infer those ideas.[76]

*   *   *

These communications networks are crucial elements for spaceflight. Without them, control rooms had no way to monitor their spacecraft. The networks have also played an important role in another aspect of control center operations. With remote sites and ground stations throughout the world, they aid in the space agencies' international relations. Finally, the communications networks provide another useful example of the similarities of control centers based on function over any other aspect, even overriding nationality.

CONCLUSION

Since the beginning of space travel, Mission Control has served as the vital link between spacecraft and the earth. All spacecraft, whether robotic or human, need the men and women on the ground monitoring their progress and maintaining their systems. Whether monitoring a routine flight or fixing a catastrophic problem, flight controllers remain alert and ready to perform their duties.

Though they host a key component of space travel, the control centers have had moments of doubt about their continued existence and have adapted out of necessity, in order to stay relevant. Flexibility remains key to the success of the control rooms. The Space Flight Operations Facility now operates as the most prominent American control room with the Mars Science Laboratory on the red planet. The Flight Control Rooms, however, in stark contrast, sit largely idle. One monitors the International Space Station while the other undergoes renovations in preparation for future missions after the retirement of the space shuttles. ESOC continues to thrive as the most active European control center, handling missions from various other space programs.

While the space agencies constructed their control centers independently and in distinct situations, they have taken on noteworthy similarities. Each has a main control room or rooms with surrounding support areas. Each has developed control systems for their computers or consoles for their particular needs. Each has a larger infrastructure to support the control rooms, including a communications network to transmit signals to and from the spacecraft. The communications networks in each case have served as aspects of international relations for their parent space agencies, particularly during the Cold War. Each control room has expert, well-trained, and highly professional controllers for various systems, with a form of team leader to oversee all the work. Simulations train the controllers and help prepare them for possible anomalies. Accidents do happen, and controllers in each control center have worked to fix those problems and rescue missions that would fail without their diligence. Control rooms remain the most prominent aspect of spaceflight, whether human or robotic, American or European.

Through this study, it became clear that the greatest differences in operation arose out of the missions monitored in the room more than anything else, including national ties. At JSC, the center for American human spaceflight, controllers must constantly monitor astronauts while they are in space. Mistakes can be the difference between life and death. At JPL and ESOC, robotic missions tend to last longer, with many lasting years or decades. To work most efficiently and economically, the main control rooms are used typically for vital aspects of missions, such as launch, rendezvous, and landing. During other times, missions are monitored by a skeleton crew of controllers in separate dedicated control rooms.

Each of the control rooms has acquired new technologies, albeit sometimes reluctantly, that have been vital to accomplishing their missions. At JSC, many of the original controllers had little experience with computers, which led JSC to recruit heavily from college campuses. Later controllers were so proficient they brought personal computers into the Mission Operation Control Rooms to supplement the mainframes, demonstrating their flexibility and contributing to the move toward the new FCRs. The new control rooms utilize off-the-shelf computers and even have software to monitor spacecraft from outside the FCRs.

As one ESA employee put it, everything in Europe is older than in America, except the space programs.[1] Throughout much of their histories, technology and information exchange moved one way: east across the Atlantic. Now NASA understands that it can learn from a smaller space program that has accomplished its goals despite having budgets and employee numbers a fraction of NASA's.[2] ESA has long automated many of its processes, particularly the remote ESTRACK stations. Now JPL has begun to ponder a move in that direction, and it looks to ESOC for guidance. This may be the most direct example of technology transfer between Mission Control Centers in this brief history.

The degree of cooperation between control rooms changed dramatically over time. While ESA always sought extensive international cooperation, NASA remained largely an independent agency during the Cold War. Reluctance to share technology, which could eventually get into the hands of their enemy, played an important role in this mindset. As the Cold War marched on, but budgets diminished, NASA was forced to seek more outside help in the 1970s and 1980s. Following the Cold War, NASA has almost become as dependent on international help as ESA to complete its missions, such as the International Space Station. Space agencies will continue to seek more cooperation with each other as budgets shrink and public expectations rise.

*　　*　　*

Flight directors of JSC have unique power when dealing with American human-spaceflight missions. Controllers and astronauts provide insight, but the FD makes the final decision. The other control rooms have positions similar to the flight director, but without the same degree of autonomy. In general, they make decisions with the consent of other controllers or with deference to other experts.

It is true that technological advances have allowed for more autonomy in onboard computers. The actual control by operatives on the ground has diminished with each new generation of software and hardware. A presence on the ground remains necessary, nevertheless, for both human and robotic missions. In each case, engineers on the ground are best suited to fix anomalies that inevitably occur in the hazards of spaceflight. As Chris Kraft, John Hodge, and Gene Kranz have said, ground control is necessary because spacecraft are pushed to extremes. Flight controllers are primarily there to "monitor, evaluate, recommend—if necessary—command" the vehicles.[3] Controllers want to be hands off as much as possible and only take control if necessary.

With the ISS, controllers on the ground monitor space station systems, thereby freeing up the astronauts to do other work. Without the assistance of the ground, ISS inhabitants did not even have enough time to take care of the station, let alone accomplish scientific goals. The space agencies have yet to develop the technology needed to fully automate spacecraft. While the space agencies are working toward less dependence on Mission Control, it remains a necessary aspect of all spaceflight. As one ESA general report stated, operators are the shepherds, satellites the sheep.[4]

* * *

Ultimately, one of the most intriguing aspects of this study is the revelation that the control centers did develop independently. Typically in history, technologies were developed in one or maybe a handful of places and then transferred from there to other locations. An excellent example was the smelting of iron, which most historians believe was only discovered once, in China, and the knowledge diffused from there to the rest of the world. A more recent example was the Manhattan Project, from which nearly all nuclear weapon technologies have arisen.

These Mission Control Centers, however, were constructed by their agencies and developed their technologies and their mode of control completely independently at nearly the same time—that is, in parallel construction. The only instance of technology transfer uncovered is that of automated network sites, which has yet to be fully implemented. Other than that, even when the centers cooperated with each other, there is no evidence of sharing technology or other ideas of control.

Despite this complete independence, the control centers still exhibit remarkable similarities. Even more intriguing, they seem to be growing more similar as cooperation increases. Among the three, Johnson Space Center has always been the outlier, since the MOCRs had little more than cosmetic similarities to the other two. As the European veteran Wolfgang Wimmer asserted, the new FCRs have transitioned JSC to a system more similar to the other two with greater flexibility. There is no recorded evidence, however, that JSC specifically made that change due to interactions with other control centers.

Perhaps the most important understanding to take away from these observations is that, regardless of the specifics, there are some general aspects necessary for Mission Control. Foremost among those is a centralized control room for critical aspects of missions, where experts can congregate in case of an emergency. Some semblance of hierarchy and authority is necessary. The systems, hardware, software, and the controllers themselves must remain flexible to adapt to changes in missions. It is evident that, as long as spaceflight ventures stay within near proximity to earth, Mission Control and the flight controllers will remain a vital aspect of space exploration.

*   *   *

It must be remembered that much of the history of the control centers occurred within the context of the Cold War. This affected nearly every aspect of their work, from the missions to be controlled to potential allies to the development and distribution of technology. The controllers of JSC, JPL, and ESOC maintained at least friendly relationships with each other, especially when compared to their interactions with controllers from the Soviet Union.

Mission Control played an integral role in dissuading fears in the public about the dangers of technology. Here were clear examples of modern technologies being used for good. Beginning with the Apollo program, spacecraft were seen as "winning" the Cold War for the United States and her allies. When American men set foot on the moon, there could be no doubt that NASA had won at least one aspect of the Cold War against the ideologically opposed Communists.

All the while, engineers and scientists were controlling the action from the ground. Potentially volatile technologies, many of which were paired with the potential end-of-world scenarios envisioned with nuclear warfare, were managed in some ways by Mission Control. Even if borne subconsciously, using the word *control* helped to suppress very real Cold War anxieties.

Finally, these elements lead to one more conclusion: the space programs should be viewed as another of the "proxy" wars in the Cold War. Instead of the superpowers fighting each other through the Vietcong or the Mujahedeen, they battled for technological supremacy in space. The term *space race* may have fallen out of favor among historians largely because the Soviet Union never successfully launched their N1 moon rocket. That does not mean, however, that space exploration in general was not an integral element of their efforts to dominate the world. There may not have been "shots fired," but there were unfortunate casualties on both sides. Like the other hot wars, the respective governments poured a huge amount of money and resources into their space programs to defeat the other world power.

While the most glamorous aspects of this proxy war occurred in space, the majority of the work transpired on the ground. As always, Mission Control was a key, indispensable element of that effort. The control centers managed the computer systems, maintained the communication networks, and oversaw all flight operations. The Mission Control Centers maintained communication between the ground to space during the Cold War and beyond.

A statement by Chris Kraft specifically about the flight controllers of the MOCR pays homage to the significance of the men and women of all of the Mission Controls.

> My flight controllers are too often unsung heroes. No mission then or now could be flown without the dedication, professionalism, and raw intelligence of the men and women who work the consoles. They are an American treasure.[5]

Mission Control is a unique space. Through those rooms, dreams have become reality. With all of their differences and similarities, they stand as the vital link between humans on earth, and the vast reaches of the heavens.

ACKNOWLEDGMENTS

The process of making this work into a complete manuscript has seen me go through some rather drastic changes, from graduate student to part-time professor trying to scrape together various jobs to full-time visiting professor to seminarian for the Roman Catholic Archdiocese of Galveston-Houston. Needless to say, this has given me the grace of working with numerous individuals without whom this book would not have been possible.

Starting at the beginning, I would like to thank the professors at Auburn University who were instrumental in getting this idea "off the ground." In particular, I thank James Hansen, Bill Trimble, and Ralph Kingston. Your support, observations, and steadying voice were just what I needed. I also need to thank the History of Science Society and NASA for their History of Space Science Fellowship. Without that financial support, my research trip to Europe would have been nearly impossible, making this final work that much more difficult.

On that and all of my research trips I was graced with the opportunity to meet exceptional individuals both in the archives and working the mission controls of NASA and ESA. While it is not possible to name all of them, there are a few I need to single out. At the Jet Propulsion Laboratory, Jim McClure and Ron Sharp were willing to give me a tour of the SFOF and a slideshow with images that proved to be the most important aspects of my research there. At ESA headquarters, Nathalie Tinjod and Mélanie Legru went above and beyond to find the few documents that had survived the years of archive culls. At the European Space Operations Centre Manfred Warhaut took time out of his busy schedule to help this American interested in the history of his mission control.

Thanks also goes to the students and professors at the universities where I have had the privilege of teaching. A special recognition goes to Bill Morison and Carolyn Shapiro-Shapin at Grand Valley State University. Without your last-minute call I would have missed out on the chance to live my teaching dream, not to mention the winter to end all winters.

I have had a significant amount of spiritual guidance along the way as well. In particular Fr. Bill Skoneki in Auburn, Fr. Mike Alber in Grand Rapids, Fr. Norbert Maduzia and Fr. James Burkart in Houston, and the members of the vocations office in Galveston-Houston have helped in numerous ways. I would also like to recognize the faculty, staff, and seminarians of St. Mary's Seminary in Houston, where I have started the next step in my journey.

What is a man without friends? Instead of mentioning everyone and inevitably forgetting someone, a certain few with great personal significance include Karen Gibbs, Cody Smith, Josh Barronton, Abby Sayers, and Adrianne Hodgin Bruce. One way or another you have been enormously instrumental in this process. Thank you from the bottom of my heart.

Finally, my family has always been tremendously important to me. Chris, Heather, Sydney, Rob, Carrie, Cooper, Caroline, Julia, and Joseph, you have always been there for me exactly as I needed you. Most important, my parents, David and Eileen, who have struggled with me, laughed with me, and inspired me these many years: I cannot thank you enough, so I will just try to be the best son I can be.

NOTES

INTRODUCTION

1. Murray and Cox, *Apollo*, 1989, 361–69; Chaikin, *A Man on the Moon, Vol. II*, 1994, 20–23; Conrad and Klausner, *Rocket Man*, 2005, 166–70.

2. This is not an unusual tactic. In the movie *Black Hawk Down*, to give just one example, a few of the soldiers are composites to make the story easier to follow.

3. Interestingly, the outside of the Mission Control Center, best seen in the eighth episode, featuring the *Apollo 13* mission, clearly is incorrect. The fictional buildings are made of brick and there is a low wall outside with a sign demarcating the Mission Control Complex.

4. Levin, "Contexts of Control," in Levin, ed., *Cultures of Control*, 2000, 21–27.

5. Williams, "Nature Out of Control," in Levin, ed., *Cultures of Control*, 2000, 56.

6. An example comes from the fabled *Apollo 13* mission. As carbon dioxide levels rose, controllers needed to organize a solution to replace the scrubbers using only the objects found onboard the spacecraft.

7. Mindell, "Beasts and Systems," in Levin, ed., *Cultures of Control*, 220; and Bowles, "Liquifying Information," in Levin, ed., *Cultures of Control*, 226–27.

8. For a more detailed account of technophobia in science fiction, please consult Dinello, *Technophobia*, 2005.

CHAPTER 1. JOHNSON SPACE CENTER

1. Kraft, *Flight: My Life in Mission Control*, 2001, 66.

2. Kraft, *Flight*, 87–93; and Murray and Cox, *Apollo*, 245–47.

3. Kranz, *Failure Is Not an Option*, 2000, 12.

4. Kraft, *Flight*, 2.

5. Kranz, *Failure Is Not an Option*, 17.

6. Eugene F. Kranz, interview by Roy Neal, 19 March 1998, transcript, John Space Center Oral History Collection, 5.

7. Ibid., 8–11.

8. Dethloff, *Suddenly, Tomorrow Came: A History of the Johnson Space Center*, 1993, 38.

9. Ibid., 39.

10. Ibid., 33, 40.

11. Murray and Cox, *Apollo*, 245.

12. James Webb, interview, 15 March 1985, transcript, Glennan-Webb-Seamans Project Interviews, National Air and Space Museum Archives, Suitland, MD, 109–10.

13. Kraft, *Flight*, 149–50.

14. Dethloff, *Suddenly Tomorrow Came*, 41–42; Kraft, *Flight*, 172.

15. Dethloff, *Suddenly Tomorrow Came*, 42–43, 45–46. Dethloff also provides great detail about Grace Winn, who managed much of the welcome for STG members moving to Houston.

16. Robert C. Seamens Jr., "Location of Mission Control Center," Memorandum for Administrator, 10 July 1962, MSC-Mission Control Center 4526, National Aeronautics and Space Administration Headquarters Archives, Washington, DC.

17. "NASA Mission Control Center at Houston," NASA News Release, No. 62-172, 20 July 1962, MSC-Mission Control Center 4712, National Aeronautics and Space Administration Headquarters Archives, Washington, DC.

18. Kraft, *Flight*, 214–15.

19. Clark, "New NASA Center Making Its Debut," 1965, 21.

20. Glynn S. Lunney, interview by Carol Butler, 28 January 1999, transcript, JSC Oral History Collection, 18–19.

21. Dethloff, *Suddenly Tomorrow Came*, 214.

22. M. P. "Pete" Frank III, interview by Doyle McDonald, 19 August 1997, transcript, JSC Oral History Collection, 23–24.

23. Dethloff, *Suddenly Tomorrow Came*, 249.

24. "Neutral Buoyancy Laboratory," National Aeronautics and Space Administration, http://dx12.jsc.nasa.gov/site/index.shtml (accessed 17 January 2012).

25. Christopher C. Kraft and Sigurd Sjoberg, "Gemini Mission Support," Gemini Mid-Program Conference, 23–25 February 1966; Kraft, Christopher: Biographical Data 1238, National Aeronautics and Space Administration Headquarters Archives, Washington, DC.

26. NASA News Release, MSC 64-8, 11 January 1964, MSC-Mission Control

Center 4712, National Aeronautics and Space Administration Headquarters Archives, Washington, DC.

27. While touring the MCC with another NASA employee, we were grateful for our knowledgeable guide, Terry Hartman. Without him we would never have found our way. The only thing I can liken the MCC corridors to is the interior of an aircraft carrier, where a visitor must have someone to show them around or risk being lost, like Charlie on the MTA.

28. "MCC Mission Control Center," undated, box 4, Mission Control Center and Real-Time Computer Complex, Center Series, Johnson Space Center History Collection at the University of Houston–Clear Lake.

29. "NASA Mission Control Center Historical Overview," 30 March 1994, MSC-Mission Control Center 4712, National Aeronautics and Space Administration Headquarters Archives, Washington, DC.

30. Kearney, "The Evolution of the Mission Control Center," 1987, 399–400.

31. Robert D. Legler, Responses to Questions about Historical Mission Control, 7 April 1997, National Aeronautics and Space Administration Headquarters Archives, Washington, DC.

32. "Spacelab Payload Control," NASA Fact Sheet, 1983, MSC-Mission Control Center 4712, National Aeronautics and Space Administration Headquarters Archives, Washington, DC.

33. "Mission Control Center," NASA Facts, August 1993, MSC-Mission Control Center 4712, National Aeronautics and Space Administration Headquarters Archives, Washington, DC.

34. The backup moved to White Sands in the 1980s. Previously it had been housed at Goddard Space Flight Center. ("Mission Control Center," NASA Facts, 1986, MSC-Mission Control Center 4712, National Aeronautics and Space Administration Headquarters Archives, Washington, DC; and Barbara Selby and Carolynne White, "Shuttle Emergency Mission Control Center Moves to White Sands," NASA News Release 88–101, 15 July 1988, MSC-Mission Control Center 4712, National Aeronautics and Space Administration Headquarters Archives, Washington, DC.)

35. "Philco Houston Mission Control Center Press Tour," undated, box 1, Mission Control Center and Real-Time Computer Complex, Center Series, Johnson Space Center History Collection at the University of Houston–Clear Lake, 2–3.

36. 1 gigabyte=1,048,576 kilobytes.

37. Kearney, "The Evolution of the Mission Control Center," 400.

38. "IBM Tour Manned Spaceflight Control Center," 10 January 1965, box 2, Mission Control Center and Real-Time Computer Complex, Center Series, Johnson Space Center History Collection at the University of Houston–Clear Lake, 3.

39. James Stroup, "Ground Computer Complex Procurement Plan," 15 August 1962, box 2, Mission Control Center and Real-Time Computer Complex, Center Series, Johnson Space Center History Collection at the University of Houston–Clear Lake.

40. "Selection of Contract for the Ground Computer Complex at the Integrated Mission Control Center," 1963, James C. Elms Collection, folder 5, box 1, Archives Division, Smithsonian National Air and Space Museum Archives, Suitland, MD; "Contract Signed with IBM for Computer Equipment," NASA News Release 63–151, 12 July 1963, MSC-Mission Control Center 4712, National Aeronautics and Space Administration Headquarters Archives, Washington, DC.; and Kraft, *Flight*, 192–93.

41. Kearney, "The Evolution of the Mission Control Center," 402; and "MCC Development History," compiled by Ray Loree, August 1990, box 1, Mission Control Center and Real-Time Computer Complex, Center Series, Johnson Space Center History Collection at the University of Houston–Clear Lake, A-2.

42. Other bidders included Amdahl Corp. and ViON Corp. (Kenneth C. Atchison and Terry White, "NASA Selects IBM to Provide Mission Control Computers," NASA News, 29 October 1985, MSC-Mission Control Center 4712, National Aeronautics and Space Administration Headquarters Archives, Washington, DC; Kearney, "The Evolution of the Mission Control Center," 405; and "MCC Development History," A-2.

43. "Old, New Meet in Mission Control," *Countdown*, July/August 1995, MSC-Mission Control Center 4712, National Aeronautics and Space Administration Headquarters Archives, Washington, DC, 29.

44. T. Rodney Loe, interview by Carol L. Butler, 30 November 2001, transcript, JSC Oral History Collection, 1.

45. William Harwood, "NASA's New Control Center to Manage February Launch," Space News, 4–10 December 1995, MSC-Mission Control Center 4712, National Aeronautics and Space Administration Headquarters Archives, Washington, DC, 7.

46. William Harwood, "'Houston' of Space Flight History Catches Up With Look of Future," *Washington Post*, 16 July 1995, MSC-Mission Control Center 4712, National Aeronautics and Space Administration Headquarters Archives, Washington, DC, A3.

47. M. P. Frank, "Flight Control of the Apollo Lunar-Landing Mission," 25 August 1969, Johnson Space Center History Collection at the University of Houston–Clear Lake, 4.

48. Murray and Cox, *Apollo*, 348–50.

49. T. Rodney Loe, interview by Carol L. Butler, 7 November 2001, transcript, JSC Oral History Collection, 53.

50. Loe, interview, 30 November 2001, 10–11.

51. Murray and Cox, *Apollo*, 350–51.

52. Gene Kranz, interview by Jo Jeffrey Kluger, 29 May 1992, Kranz, Eugene F. (NASA-Bio.) 1243, National Aeronautics and Space Administration Headquarters Archives, Washington, DC.

53. Paul Purser, interview by Robert Merrifield, 17 May 1967, transcript, Manned Spacecraft Center (MSC) History Interviews Kr-Z, folder 15, box 2, 18994, National Aeronautics and Space Administration Headquarters Archives, Washington, DC, 17.

54. Kranz, *Failure Is Not an Option*, 125–26. FIDO Ed Pavelka contradicts this story. He states that while the Houston controllers had better information than those at Mercury Control, primary control was never handed over. Edward L. Pavelka Jr., interview by Carol Butler, 26 April 2001, transcript, JSC Oral History Collection, 8–9.

55. Kranz, *Failure Is Not an Option*, 204–5.

56. Kranz, interview, 19 March 1998, 17.

57. Gene Kranz and others, Series of Emails about "Clear the Tower," MSC-Mission Control Center 4712, 29 June 2007.

58. NASA News Release, MSC 64-8.

59. Kraft, *Flight*, 218–19.

60. NASA News Release, MSC 64-8; Kraft, *Flight*, 193; and "Contractual History of Major Implementations and Operations Milestones," 10 January 1985, box 1, Mission Control Center and Real-Time Computer Complex, Center Series, Johnson Space Center History Collection at the University of Houston–Clear Lake.

61. James Atwater, "The Men Who Control Our Missions to the Moon," *Saturday Evening Post* (28 December 1968/11 January 1969), MSC-Mission Control Center 4712, National Aeronautics and Space Administration Headquarters Archives, Washington, DC, 36.

62. Marianne J. Dyson, "Shuttle Mission Control," JSC Shuttle Mission Control (1981–1991) 007095, National Aeronautics and Space Administration Headquarters Archives, Washington, DC, 3–5.

63. Arnold D. Aldrich, interview by Kevin M. Rusnak, 24 June 2000, transcript, JSC Oral History Collection, 47.

64. John D. Hodge, interview by Rebecca Wright, 18 April 1999, transcript, JSC Oral History Collection, 25.

65. Dyson, "Shuttle Mission Control," 5.

66. "MCC Mission Control Center," undated.

67. Atwater, "The Men Who Control Our Missions to the Moon," 69.

68. Eugene F. Kranz, interview by Rebecca Wright, 8 January 1999, transcript, JSC Oral History Collection, 55–56.

69. John Aaron, interview by Kevin M. Rusnak, 26 January 2000, transcript, JSC Oral History Collection, 7.

70. Dyson, "Shuttle Mission Control," 5–6; and Charles L. Dumis, interview by Kevin M. Rusnak, 1 March 2002, transcript, JSC Oral History Collection, 40.

71. Eugene F. Kranz, interview by Roy Neal, 28 April 1999, transcript, JSC Oral History Collection, 13.

72. Michael Behar, "The Ground," *Air & Space* (October/November 2006), MSC-Mission Control 4712, National Aeronautics and Space Administration Headquarters Archives, Washington, DC, 30–31, 35.

73. "NASA Chooses Three New Flight Directors to Lead Mission Control," NASA Press Release 09-133, 12 June 2009, MSC-Mission Control Center 4712, National Aeronautics and Space Administration Headquarters Archives, Washington, DC.

74. Behar, "The Ground," 31.

75. Kraft, *Flight*, 2.

76. Kranz, interview, 19 March 1998, 10.

77. Gene Bylinsky, "When the Countdown Is 1 . . . His Pulse Is 135," *The New York Times Magazine* (15 August 1965), Kraft, Christopher: Biographical Data 1237, National Aeronautics and Space Administration Headquarters Archives, Washington, DC, 13.

78. Kranz, interview, 28 April 1999, 41.

79. "At 18 Years Old She Mans a Mission Control Console," NASA News Release 79-17, 19 March 1979, Parker, Jackie (NASA Bio.) 1658, National Aeronautics and Space Administration Headquarters Archives, Washington, DC.

80. This is not unusual, since most high-stress jobs, like air traffic control or police work, have relatively low average ages.

81. Kranz, interview, 28 April 1999, 13.

82. Kranz, *Failure Is Not an Option*, 304.

83. Anne L. Accola, interview by Rebecca Wright, 16 March 2005, transcript, JSC Oral History Collection, 5, 24.

84. Kranz, interview, 28 April 1999, 13.

85. Behar, "The Ground," 30.

86. Dwayne Brown, Sonja Alexander, and Kylie Clem, "First Hispanics on Duty Leading Mission Control Team," NASA News Release 05-411, 18 November 2005, MSC-Mission Control Center, National Aeronautics and Space Administration Headquarters Archives, Washington, DC.

87. Kranz, *Failure Is Not an Option*, 257–58.

88. Frank, interview, 19 August 1997, 15–16.

89. Kranz, interview, 8 January 1999, 55–56.

90. Jay H. Greene, interview by Sandra Johnson, 8 December 2004, transcript, JSC Oral History Collection, 8.

91. Charles Lewis, interview by author, 22 September 2006, transcript, Skylab Oral History Project, University of North Texas Oral History Program, 20–21; and Kranz, interview, 8 January 1999, 24.

92. Lewis, interview, 22 September 2006, 21.

93. Kranz, *Failure Is Not an Option*, 240–42; and Edward L. Pavelka Jr., interview by Carol Butler, 9 March 2001, transcript, JSC Oral History Collection, 1–3.

94. Edward L. Pavelka, "The Origin of Captain Refsmmat," *MOD Focus* (October 1985), Box 7A, Mission Operations, Center Series, Johnson Space Center History Collection at University of Houston–Clear Lake, 7.

95. Pavelka, interview, 26 April 2001, 14–15.

96. Jay H. Greene, interview by Sandra Johnson, 10 November 2004, transcript, JSC Oral History Collection, 11, 35.

97. Glynn S. Lunney, interview by Roy Neal, 9 March 1998, transcript, JSC Oral History Collection, 14.

98. Ken Peek, "History of Human Spaceflight Mission Operations," *Quest* 10:4 (2003), MSC-Mission Control Center 4712, National Aeronautics and Space Administration Headquarters Archives, Washington, DC, 23.

99. Dumis, interview, 1 March 2002, 16.

100. Charles S. Harlan, interview by Kevin M. Rusnak, 14 November 2001, transcript, JSC Oral History Collection, 19–20.

101. Rita Karl, "Mission Control Center Apollo," PowerPoint presentation, 18 June 2001, Johnson Space Center History Collection at the University of Houston–Clear Lake; and Dyson, "Shuttle Mission Control," 9–10, 18–19, 25–26, 32–33, 42–43, 49–50, 54–55, 62–63, 74–76, 87–88, 99–100, 108–9, 118–19, 127–28, 138–39, 149–50, 160–62. 171–72, 184–85.

102. Eugene Kranz, interview by author, 22 November 2006, transcript, Skylab Oral History Project, University of North Texas Oral History Program, 22–24.

103. James Hartsfield, "Flight Control of STS-59," NASA News Release 94-026, 29 March 1994, MSC-Mission Control Center, National Aeronautics and Space Administration Headquarters Archives, Washington, DC.

104. Holkan, "Shift Change Briefing Plan," MCC Houston Standard Operating Procedures, Manned Spaceflight Center, 10 May 1965, Record Group 255, National Archives, Fort Worth, Texas, 10-2.

105. Kranz, interview, 8 January 1999, 38.

106. Bridget Mintz Testa, "Mission Control," *Invention & Technology* 18, no. 4 (Spring 2003), MSC-Mission Control Center 4712, National Aeronautics and Space Administration Headquarters Archives, Washington, DC, 17.

107. Dyson, "Shuttle Mission Control," 14–15.

108. "MCC Development History," 15.

109. Kearney, "The Evolution of the Mission Control Center," 399.

110. "Old, New Meet in Mission Control," 29.

111. "MCC Development History," 23, A-2.

112. Testa, "Mission Control," 21.

113. Ibid., 24.

114. Dyson, "Shuttle Mission Control," 1.

115. James Hartsfield, "New Mission Control Room portends increasing pace of human space flight, rapid space station expansion," *Space Center Roundup* 39, no. 18 (8 September 2000), MSC-Mission Control Center, National Aeronautics and Space Administration Headquarters Archives, Washington, DC, 1.

116. "Old, New Meet in Mission Control," 29.

117. Annette P. Hasbrook, interview by Jennifer Ross-Nazzal, 21 July 2009, transcript, JSC Oral History Collection, 1–3, 24.

118. Testa, "Mission Control," 22.

119. Harwood, "'Houston' of Space Flight History Catches Up With Look of Future," A3.

120. Sean Wilson, "Mission Operations Evolving for Spaceflight of the Future," *Roundup*, MSC-Mission Control 4712, National Aeronautics and Space Administration Headquarters Archives, Washington, DC, 6–7.

121. "NASA Modifies Mission Operations Support Contract," HQ News C09-031, 23 June 2009, MSC-Mission Control Center 4712, National Aeronautics and Space Administration Headquarters Archives, Washington, DC; and "NASA Extends Mission Operations Support Contract," News Release C11-053, 12 December 2011, http://www.nasa.gov/home/hqnews/2011/dec/HQ_C11-053_FDOC_Contract.html.

122. Hasbrook, interview, 30–31.

123. Lipartito and Butler, *A History of the Kennedy Space Center*, 2007, 50.

124. NASA Public Affairs, *The Kennedy Space Center Story* (Cape Canaveral, FL: Kennedy Space Center, 1991), 14.

125. Lipartito and Butler, 191.

126. Lipartito and Butler, 103–4; and National Aeronautics and Space Administration Public Affairs, *The Kennedy Space Center Story*, Kennedy Space Center, FL: John F. Kennedy Space Center, 1972, 24.

127. NASA Public Affairs, *The Kennedy Space Center Story*, 23–25.

128. Lipartito and Butler, 211, 224.

129. Lipartito and Butler, 18.

130. Ibid., 20–21.

131. The *Kobayashi Maru* represents a no-win scenario, as seen in the *Star Trek* universe.

132. Kraft, *Flight*, 113–14.

133. Christopher C. Kraft Jr., "Mercury Operational Experience," 30 April–2 May 1962, Record Group 255, W66-10, 306, National Archives, Fort Worth, Texas.

134. Hodge, interview, 28.

135. Gerald D. Griffin, interview by Doug Ward, 12 March 1999, transcript, JSC Oral History Collection, 2–3.

136. Harlan, interview, 18–19.

137. Hasbrook, interview, 16–17.

138. Loe, interview, 30 November 2001, 3–4.

139. Lunney, interview, 9 March 1998, 46.

140. The audio seems to indicate that Jack Swigert was the first to indicate that they had a problem. When asked for clarification, Lovell responded with the famous phrase.

141. Chaikin, *A Man on the Moon, Vol. II*, 119.

142. There are many good sources for further details on *Apollo 13*. In particular, please see: Murray and Cox, *Apollo*, 377–434, Chaikin, *A Man on the Moon, Vol. II*, 95–172, and Jim Lovell and Jeffrey Kluger, *Lost Moon: The Perilous Voyage of Apollo 13* (1994).

143. Murray and Cox, *Apollo*, 338–46; Chaikin, *A Man on the Moon, Vol. I*, 293–98; and Hansen, *First Man: The Life of Neil A. Armstrong*, 2005), 458–75.

144. This incident also provides an example of the importance of naming technology. Officially, the shield was known simply as the micrometeoroid shield, leading to some NASA engineers' misunderstanding its importance to the station. Those who knew of its primary job, as a heat shield to block solar radiation, immediately knew the disaster that could result from losing such an essential element.

145. Thomas Y. Canby, "Skylab," 1974, 457.

146. Shayler, *Skylab: America's Space Station*, 2001, 212.

147. Paul J. Weitz, interview by Rebecca Wright, 26 March 2000, transcript, JSC Oral History Collection, 66.

148. Kranz, *Failure Is Not an Option*, 59.

149. Loe, interview, 7 November 2001, 7–8.

150. NASA Press Release, 1973, Skylab Air-to-Ground 7637, National Aeronautics and Space Administration Headquarters Archives, Washington, DC; and James C. Fletcher, "Private Communications for Skylab," Memorandum, 3 May 1973, Skylab Air-to-Ground 7637, National Aeronautics and Space Administration Headquarters Archives, Washington, DC.

CHAPTER 2. JET PROPULSION LABORATORY

1. The history of the Jet Propulsion Laboratory has been extensively discussed by the two center histories, Koppes's *JPL and the American Space Program* (1982) and Westwick's *Into the Black* (2007). The first two sections of this chapter, therefore, will strive to present a brief overview of the history of the center before delving into mission control itself. As a result, much of the information in these two sections comes from those sources.

2. Von Kármán, *The Wind and Beyond*, 1967, 234–48; and Frank J. Malina, "On The GALCIT Rocket Research Project, 1936–38," in Durant and James, *First Steps Towards Space*, 1974, 113–14.

3. Frank J. Malina, "The U.S. Army Air Corps Jet Propulsion Research Project, GALCIT Project No. 1, 1939–1946: A Memoir," in Hall, *Essays on the History of Rocketry and Astronautics,*1977, 154–58.

4. Von Kármán, *The Wind and Beyond*, 250.

5. Ibid., 254.

6. Koppes, *JPL and the American Space Program*, 44–45.

7. Ibid., 20–21.

8. Dunn, Powell, and Seifert, "Heat Transfer Studies Relating to Rocket Power Plant Development,", 1951, 271–328; and Koppes, *JPL and the American Space Program*, 23–24. V-2 rockets launched by the German rocket program during World War II reached space at least three years earlier than the first American rocket.

9. Koppes, *JPL and the American Space Program*, 25–29.

10. Koppes, *JPL and the American Space Program*, 30–32; Von Kármán, *The Wind and Beyond*, 308–15; and Viorst, "The Bitter Tea of Dr. Tsien," 1967.

11. See chapters 3 and 4 of Koppes, *JPL and the American Space Program*.

12. Ibid., 48–50.

13. Ibid., 67–70.

14. Koppes, *JPL and the American Space Program*, 85–86; and William H. Pickering and James H. Wilson, "Countdown to Space Exploration: A Memoir of the Jet Propulsion Laboratory, 1944–1958," in Hall, *Essays on the History of Rocketry and Astronautics*, 1977), 416–18.

15. Koppes, *JPL and the American Space Program*, 96–99.

16. Mudgway, *Uplink-Downlink*, 2001, 11.

17. Koppes, *JPL and the American Space Program*, 99.

18. For more on this "troubled triangle," please see Koppes, *JPL and the American Space Program*, chapter 9.

19. Koppes, *JPL and the American Space Program*, 216.

20. Ibid., 215–16.

21. "The Space Flight Operations Facility," *JPL Space Programs Summary* 37–20, vol. 6, Jet Propulsion Laboratory Library, Pasadena, California, 44.

22. Press Release, History Collection 3-170e, Jet Propulsion Laboratory Library, Pasadena, California.

23. William Pickering, "The Jet Propulsion Laboratory," *Bulletin of CIT* 73, no. 4 (19 November 1964), History Collection 3–168, Jet Propulsion Laboratory Library, Pasadena, California, 44.

24. "Radar Signals 'Christen' SFOF," *Lab-oratory*, vol. 13, no. 10, May 1964, Jet Propulsion Laboratory Library, Pasadena, California, 3.

25. Homer E. Newell, "Dedication Remarks," 14 May 1964, History Collection 3-170c, Jet Propulsion Laboratory, Pasadena, California.

26. "The Space Flight Operations Facility," *JPL Space Programs Summary* 37–20, vol. 6, 46.

27. "SFOF Near Ready," *Lab-oratory*, vol. 13, no. 3, October 1963, Jet Propulsion Laboratory Library, Pasadena, California, 3; and National Register of Historic Places Inventory-Nomination Form, United States Department of the Interior, NPS Form 10-900 (7-81), 15 May 1984, Jet Propulsion Laboratory Library, Pasadena, California.

28. Press Release, HC Docs 3-170e, 2–3.

29. Memorandum from Brian Sparks to Division Chiefs and Managers, 28 July 1964, History Collection 2–281, Jet Propulsion Laboratory Library, Pasadena, California; and "The Deep Space Network," Jet Propulsion Laboratory 1965 Annual Report, Jet Propulsion Laboratory Library, Pasadena, California, 23.

30. Memorandum from A. R. Luedecke to Senior Staff, Section Managers, etc., 5 April 1965, History Collection 2–284, Jet Propulsion Laboratory Library, Pasadena, California.

31. Letter from NASA Headquarters to JPL, History Collection 2–175, Jet Propulsion Laboratory Library, Pasadena, California.

32. Letter from NASA Headquarters to JPL.

33. "The Deep Space Network," Jet Propulsion Laboratory 1965 Annual Report, 23.

34. Mudgway, *Uplink-Downlink*, 42.

35. Press Release, HC 3-170e.

36. "The Deep Space Network," Jet Propulsion Laboratory 1965 Annual Report, 23.

37. "Space Flight Operations Facility," History Collection 3–169, Jet Propulsion Laboratory Library, Pasadena, California.

38. Press Release, HC 3-170e.

39. Press Release, HC 3-170e; and D. A. Nelson compiled, "Engineering Planning Document No. 143: Capabilities and Procedures," 20 July 1964, History Collection 2–1871, Jet Propulsion Laboratory Library, Pasadena, California, I-1.

40. "Project Viking '75 Mission Control and Computing Center Support Plan," 15 May 1975, Institutional Management Committee Collection, JPL 136, Box 8, Folder 83, Jet Propulsion Laboratory Library, Pasadena, California; and Jim McClure and Ron Sharp, interview by author, Pasadena, California, 3 September 2010.

41. Press Release, HC 3-170e.

42. "Viking'75 Project Mission Control and Computing Center System Functional Capabilities," 15 March 1973, Institutional Management Committee Collection, JPL 136, box 4, folder 27, Jet Propulsion Laboratory Library, Pasadena, California; and "Viking'75 Project Mission Control and Computing Center System Progress Review," 11 October 1973, Institutional Management Committee Collection, JPL 136, box 5, folder 38, Jet Propulsion Laboratory Library, Pasadena, California.

43. Jim McClure and Ron Sharp, interview by author, Pasadena, California, 3 September 2010.

44. André Jansson and Amanda Lagerkvist, "What Is Strange about Strange Spaces?" in Jansson and Lagerkvist, *Strange Spaces: Explorations into Mediated Obscurity*, 2009, 20–21.

45. "A New Building Goes Up," *Lab-oratory*, vol. 20, no. 5, January-February 1971, Jet Propulsion Laboratory Library, Pasadena, California, 2.

46. Memorandum from F. H. Felberg to R. J. Parks and P. T. Lyman, 1 April 1981, Institutional Management Committee Collection, JPL 228, box 2, folder 9, Jet Propulsion Laboratory Library, Pasadena, California.

47. "Scope of SFOF Transition Activities," March 1981, Institutional Management Committee Collection, JPL 228, box 2, folder 9, Jet Propulsion Laboratory Library, Pasadena, California; and Memorandum from W. J. York Jr. to J. P. Click, 20 April 1981, Institutional Management Committee Collection, JPL 228, box 2, folder 11, Jet Propulsion Laboratory Library, Pasadena, California.

48. Mudgway, *Uplink-Downlink*, xxxv.

49. "Lab Facilities Named Historic Landmarks," *JPL Universe*, vol. 24, no. 16, 12 August 1994, Jet Propulsion Laboratory Library, Pasadena, California, 3; and National Register of Historic Places Inventory-Nomination Form.

50. Corliss, *A History of the Deep Space Network*, 1976, 8.

51. Mudgway, *Uplink-Downlink*, 31.

52. Corliss, *A History of the Deep Space Network*, 38–39.

53. Ibid., 56.

54. N. A. Renzetti, K. W. Linnes, D. L. Gordon, and T. M. Taylor, "Tracking and Data System Support for the Mariners Mars 1969 Mission, Planning Phase Through Midcourse Maneuver," Technical Memorandum 33-474, Vol. I, 15 May 1971, Jet Propulsion Laboratory Library, Pasadena, California, 15.

55. N. A. Renzetti, "Tracking and Data System Support for *Surveyor* Missions I and II," Technical Memorandum 33-301, Vol. I, 15 July 1969, Jet Propulsion Laboratory Library, Pasadena, California, 158–59.

56. H. W. Alcorn, preparer, "Operations Management Plan for the Mission Control and Computing Center," 1 February 1973, Viking Project Flight Operations and Mission Control Documents, box 3, folder 24, Jet Propulsion Laboratory Library, Pasadena, California.

57. "Viking Flight Team Meeting," Mission Control Directorate Status Review, 18–21 June 1974, JPL 136, box 6, folder 48, Jet Propulsion Laboratory Library, Pasadena, California.

58. McClure and Sharp, interview.

59. "The Space Flight Operations Facility," JPL Space Programs Summary No. 37-20, Vol. 6, Jet Propulsion Laboratory Library, Pasadena, California, 61.

60. McClure and Sharp, interview.

61. Renzetti et al, "Tracking and Data System Support for the Mariners Mars 1969 Mission, Planning Phase Through Midcourse Maneuver," 46, 99–101.

62. McClure and Sharp, interview.

63. Since much of this renovation information could not be located, much of it is derived from a series of photographs in a slideshow history created by Jim McClure. This slideshow was generously shared with the author by Mr. McClure.

64. McClure and Sharp, interview.

65. Ibid.

66. Ibid.

67. Ibid.

68. Ibid.

69. Ibid.

70. Mudgway, *Uplink-Downlink*, 86.

71. McClure and Sharp, interview.

72. Ibid.

73. Renzetti et al, "Tracking and Data System Support for the Mariners Mars 1969 Mission, Planning Phase Through Midcourse Maneuver," 15.

74. Renzetti et al, "Tracking and Data System Support for the Mariners Mars 1969 Mission, Planning Phase Through Midcourse Maneuver," 139–42.

75. Bergreen, *Voyage to Mars*, 2000, 81–82.

76. Mishkin, *Sojourner: An Insider's View of the Mars Pathfinder Mission*, 2004, 76–79.

77. Bergreen, *Voyage to Mars*, 4, 75, 252–55.

78. Bergreen, *Voyage to Mars*, 81; and Mishkin, *Sojourner*, 127.

79. Bergreen, *Voyage to Mars*, 71–81. Also see Mishkin, *Sojourner*, 76–79, 92, 106, and 131–33.

80. Dethloff and Schorn, *Voyager's Grand Tour*, 2003, 138–41; and Westwick, 186–88.

81. Bergreen, *Voyage to Mars*, 86–87.

82. Webster and Brown, "NASA's Mars Rover Spirit," 2010.

83. Webster, "NASA's Spirit Rover Completes Mission on Mars," 2011; and Webster and Brown, "NASA Concludes Attempts to Contact Mars Rover Spirit," 2011.

CHAPTER 3. EUROPEAN SPACE OPERATIONS CENTRE

1. Leo Hennessy, interview by author, Paris, France, 12 October 2010.

2. Krige and Russo, *A History of the European Space Agency, Volume I*, 2000, 3.

3. Numerous histories speak to this phenomenon, including Kristin Ross, *Fast Cars, Clean Bodies: Decolonization and the Reordering of French Culture*

(1995); Gabrielle Hecht, *The Radiance of France: Nuclear Power and National Identity after World War II* (1998); Hanna Schissler, ed., *The Miracle Years: A Cultural History of West Germany, 1949–1968* (2001); Arthur Marwick, *The Sixties: Cultural Revolution in Britain, France, Italy and the United States, c. 1958–c. 1974* (1998); and Paul Betts, *The Authority of Everyday Objects: A Cultural History of West German Industrial Design* (2004).

4. CERN's twelve founding member nations were Belgium, Denmark, France, Greece, Italy, the Netherlands, Norway, Sweden, Switzerland, the United Kingdom, West Germany, and Yugoslavia.

5. Current members include those listed previously, excluding Yugoslavia and including the unified Germany, along with Austria, Bulgaria, the Czech Republic, Finland, Hungary, Poland, Portugal, Slovakia, and Spain. Romania will join in 2015. Observers include India, Israel, Japan, Russia, Turkey, and the United States, as well as the European Commission and UNESCO.

6. Krige and Russo, *A History of the European Space Agency, Volume I*, 48–51.

7. It should be noted that while ESRO, and later ESA, has many of the same member states as the European Union, they do not completely overlap.

8. Bondi, "International Cooperation in Space," 1971, 3; and Krige and Russo, *A History of the European Space Agency, Volume I*, 11.

9. Krige and Russo, *A History of the European Space Agency, Volume I*, 36–37.

10. *ESA Annual Report 1975* (European Space Agency, 1976), 111; and "Convention for the Establishment of a European Space Agency," European Space Operations Centre Library, Darmstadt, Germany, 52.

11. "Convention for the Establishment," 53.

12. Ibid., 60.

13. "Convention for the Establishment," 54; Manfred Warhaut, "ESA and ESOC Overview" (PowerPoint presentation, European Space Operations Centre, Darmstadt, Germany, 16 August 2010); and "European Space Operations Directorate: . . . constant vigil," BR 88, European Space Agency Headquarters Library, Paris.

14. "Convention for the Establishment," 58–59.

15. Ibid., 60.

16. Black and Andrews, *ESOC Services Catalog 2000/2001*, 2000, 4.

17. Gibbons, *The European Space Agency*, 1986, 21.

18. Krige and Russo, *A History of the European Space Agency, Volume I*, 53–54.

19. First General Report of the European Space Research Organization (1964–1965), European Space Operations Centre Library, Darmstadt, Germany, III.1, 1.

20. Ibid., I.6, 1.

21. Ibid., III.4, 1–4.

22. First General Report of the European Research Space Organization, I.6, 1.

23. Ibid., III.6, 1–4.

24. Bonnet and Manno, *International Cooperation in Space*, 1994, 8.

25. Schäfer, *How to Survive in Space, Volume I*, 1997, 1.

26. "The European Space Data Centre—An Establishment of the European Space Research Organization," ESRO 6380, European University Institute Archives, Florence, Italy, 2–3.

27. First General Report of the European Research Space Organization, III.6, 1–4.

28. Nye, *ESOC: European Space Operations Centre*, 1996, 6.

29. ESRO General Report 1966, European Space Agency Headquarters Library, Paris, 80; and First General Report of the European Space Research Organization, Fig. 4.1.

30. ESOC Monthly Report, December 1969, ESRO 6979, European University Institute Archives, Florence, Italy; and ESOC Bi-Monthly Report for November/December 1970, ESRO 6981, European University Institute Archives, Florence, Italy.

31. Warhaut, "ESA and ESOC Overview."

32. Longdon and David, compiled and ed., *ESOC: The European Space Operations Centre*, 1988.

33. European Space Agency Bulletin, No. 1, June 1975, European Space Agency Headquarters Library, Paris, 9.

34. Nye, *ESOC*, 2.

35. Krige and Russo, *A History of the European Space Agency, Volume 1*, 384; and "Draft of Building Lease," 22 February 1963, COPERS 45, European University Institute Archives, Florence, Italy.

36. "Contract for the Establishment of an 'Erbbaurecht' (Heritable Building Right)," 14 September 1971, ESRO 6378, European University Institute Archives, Florence, Italy, 2–5.

37. "Note on the Main Points Discussed during a Meeting Held on 11 March 1964 in Darmstadt," 19 March 1964, COPERS 45, European University Institute Archives, Florence, Italy, 2–3.

38. Professor Auger, "Extension of the Present Premises of the ESRO Establishment in Darmstadt," Memorandum to German Delegation, 27 July 1967, ESRO 6379, European University Institute Archives, Florence, Italy; and "Letter to Professor Bondi from Lindner," 24 October 1969, ESRO 6379, European University Institute Archives, Florence, Italy.

39. First General Report of the European Space Research Organization, III.6, 1–4; European Space Research Organization General Report 1966, 79; and "Report on Recent Activity at ESDAC," 8 February 1967, ESRO 6979, European University Institute Archives, Florence, Italy.

40. European Space Research Organization General Report 1966, 78; Longdon and Guyenne, *Twenty Years of European Cooperation in Space*, 1984, 88; and Krige and Russo, *A History of the European Space Agency, Volume 1*, 384.

41. Norman Longdon, ed., "ESA/ESOC 25 Years," ESA BR-90, 1992, European Space Agency Library, Paris.

42. Speech by Harry B. Gould, Site Engineer of ESDAC, 12 November 1965, European University Institute Archives, Florence, Italy.

43. J. A. Jensen, "Milestones in ESOC's History," ESA Bulletin 7 (November 1976): 56.

44. European Space Research Organization General Report 1967, European Space Agency Library, Paris, 89.

45. John Noyes, "ESOC's Participation in the Aurorae Mission," ESRO/ELDO Bulletin (April 1969 Supplement): 35.

46. "Minutes of the Meeting held on 8/1/68 at ESOC, Darmstadt, on the Use of ESOC Equipment," ESRO 6912, European University Institute Archives, Florence, Italy.

47. European Space Research Organization General Report 1969, European Space Agency Headquarters Library, Paris, 141; and "Justification for a New Control Centre Operation Building at Darmstadt," 1968(?), ESRO 6506, European University Institute Archives, Florence, Italy.

48. Longdon and Guyenne, Twenty Years of European Cooperation in Space, 92; and European Space Research Organization General Report 1969, 141.

49. European Research Organization Council, "ESOC Structure," 8 February 1971, ESRO/C (71) 9, European University Institute Archives, Florence, Italy.

50. European Space Research Organization General Report 1973, European Space Agency Headquarters Library, Paris, 127–29.

51. Schäfer, How to Survive in Space, Vol. I, 109.

52. ESA Annual Report 1990, European Space Agency Headquarters Library, Paris, 176; ESA Annual Report 1991, European Space Agency Headquarters Library, Paris, 168; and ESA Annual Report 1995, European Space Agency Headquarters Library, Paris, 121.

53. Schäfer, How to Survive in Space, Volume II, 137.

54. For more information on ISO, consult their detailed website: International Organization for Standardization, http://www.iso.org/iso/home.html.

55. ESA Annual Report 1999, European Space Agency Headquarters Library, Paris, 51.

56. Schäfer, How to Survive in Space, Vol. I, iii.

57. Wolfgang Wimmer, interview by author, Darmstadt, Germany, 21 October 2010.

58. Longdon, "ESA/ESOC 25 Years."

59. Comments by Dr. Gerhard Bengeser, Assistant Director for Administration of ESDAC, 12 November 1965, ESRO 6380, European University Institute Archives, Florence, Italy.

60. Wolfgang Hell, interview by author, Darmstadt, Germany, 20 October 2010. Hell specifically mentioned problems when moving to Italy to work at

ESRIN, but the same ideas have been expressed elsewhere with ESOC and the other centers.

61. Schäfer, *How to Survive in Space, Vol. 1*, 30–31; and Hennessy, interview.

62. Longdon, "ESA/ESOC 25 Years."

63. Schäfer, *How to Survive in Space, Vol. II*, 107; and Longdon, "ESA/ESOC 25 Years." Please note: the pamphlet refers to the sport as soccer, not football as might be expected.

64. Schäfer, *How to Survive in Space, Vol. I*, 31.

65. Schäfer, *How to Survive in Space, Vol. II*, 9–11.

66. European Space Research Organization General Report 1969, 84.

67. Warhaut, "ESA and ESOC Overview."

68. Krige and Russo, *A History of the European Space Agency, Volume I*, 72.

69. Gibbons, *The European Space Agency*, 33.

70. Bonnet and Manno, *International Cooperation in Space*, 49–50.

71. Hell, interview.

72. Bonnet and Manno, *International Cooperation in Space*, 48–49.

73. Schäfer, *How to Survive in Space, Vol. I*, 33–34.

74. Wimmer, interview. Wimmer was a contractor from 1965 to 1970 and an ESOC staff member from 1970 to 2004. Other examples abound.

75. Bonnet and Manno, *International Cooperation in Space*, 53.

76. Longdon and Guyenne, *Twenty Years of European Cooperation in Space*, 92–93.

77. European Space Research Organization General Report 1969, 141.

78. LEOP is also sometimes described as launch and early orbit.

79. Warhaut, interview.

80. "Justification for a New Control Centre Operation Building at Darmstadt."

81. Longdon and Guyenne, *Twenty Years of European Cooperation in Space*, 93.

82. European Space Research Organization General Report 1969, 141.

83. Longdon and Guyenne, *Twenty Years of European Cooperation in Space*, 92; and European Space Research Organization General Report 1969, 141.

84. Longdon and Guyenne, *Twenty Years of European Cooperation in Space*, 91.

85. European Space Research Organization General Report 1971, part 1, European Space Agency Headquarters Library, Paris, 65.

86. European Space Research Organization General Report 1972, European Space Agency Headquarters Library, Paris, 191.

87. European Space Research Organization General Report 1972, 191–92.

88. European Space Research Organization General Report 1973, 123.

89. Longdon and Guyenne, *Twenty Years of European Cooperation in Space*, 94.

90. Wimmer, interview.

91. European Space Research Organization General Report 1973, 123–24.

92. European Space Research Organization General Report 1974, European Space Agency Headquarters Library, Paris, 147; and D.E.B. Wilkins, "Spacecraft Organizations at ESOC," *ESA Bulletin* 20 (November 1979): 4.

93. Longdon and Guyenne, *Twenty Years of European Cooperation in Space*, 97.

94. "ESA's Medium Term Plan for Computer Facilities," 11 April 1983, ESA 7208, European University Institute Archives, Florence, Italy, 3.

95. European Space Agency Annual Report 1975, European Space Agency Headquarters Library, Paris, 118–19.

96. European Space Agency Annual Report 1977, European Space Agency Headquarters Library, Paris, 150–53.

97. "ESA's Medium Term Plan for Computer Facilities," 1.

98. European Space Agency Annual Report 1980, European Space Agency Headquarters Library, Paris, 114; and "ESA's Medium Term Plan for Computer Facilities," 3.

99. "ESA's Medium Term Plan for Computer Facilities," 2.

100. European Space Agency Annual Report 1981, European Space Agency Headquarters Library, Paris, 119–21.

101. Longdon and Guyenne, *Twenty Years of European Cooperation in Space*, 98.

102. "ESA's Medium Term Plan for Computer Facilities," 3; and D. Wilkins, "The European Space Operations Centre's New Control Centre," *ESA Bulletin* 43 (August 1985): 24–25.

103. Schäfer, *How to Survive in Space, Vol. 1*, 131, 133.

104. Wilkins, "The European Space Operations Centre's New Control Centre," 24–26.

105. European Space Agency Annual Report 1986, European Space Agency Headquarters Library, Paris, 175.

106. European Space Agency Annual Report 1990, 136.

107. K. Debatin, "New Ground Data-Processing System to Support the Agency's Future Satellite Missions," *ESA Bulletin* 53 (February 1988): 76–77.

108. Schäfer, *How to Survive in Space, Vol. 2*, 17–20.

109. Longdon, "ESA/ESOC 25 Years."

110. Black and Andrews, *ESOC Services Catalog 2000/2001*, 4.

111. Longdon and David, *ESOC: The European Space Operations Centre*.

112. L. Marelli and G. Valentiny, "The Control Centre and Spacecraft Control," *ESA Bulletin* 7 (November 1976): 14.

113. Nye, *ESOC: European Space Operations Centre*, 13.

114. Jocelyne Landeau-Constantin, Bernhard von Weyhe, and Nicola Cebers de Sousa, compilers, *ESOC* (Noordwijk, The Netherlands: ESA Publications Division, 2007), 36.

115. Warhaut, interview.

116. Jurgen Fertig, interview with author, Darmstadt, Germany, 20 October 2010.

117. Nye, *ESOC: European Space Operations Centre*, 17.

118. Wimmer, interview.

119. F. W. Stainer and H. P. Dworak, "Training in Satellite Ground-System Operations," *ESA Bulletin* 25 (February 1981): 68.

120. J. J. Gujer and E. Jabs, "Use of Spacecraft Simulators at ESOC," *ESA Bulletin* 59 (August 1989): 41.

121. Warhaut, interview.

122. Ibid.

123. Ibid.

124. Ibid.

125. Wilkins, "Spacecraft Operations at ESOC," 7–8.

126. Schäfer, *How to Survive in Space, Vol. 1*, 100.

127. Wilkins, "Spacecraft Operations at ESOC," 7–8.

128. Warhaut, "ESA and ESOC Overview."

129. Wimmer, interview.

130. European Space Agency Annual Report 1998, European Space Agency Headquarters Library, Paris, 97.

131. Longdon, "ESA/ESOC 25 Years."

132. H. Bath, "Operations Support," *ESA Bulletin* 2 (August 1975): 35.

133. Frank, "Flight Control of the Apollo Lunar-Landing Mission," 3–4.

134. Wimmer, interview.

135. Warhaut, interview.

136. Black and Andrews, *ESOC Services Catalog 2000/2001*, 5.

137. Wimmer, interview.

138. J. Toussaint, coordinator, "Proposal of Organisation for the Control of In-Flight Activities," 15 October 1969, ESRO 6947, European University Institute Archives, Florence, Italy, 3; and Nye, *ESOC*, 11.

139. Wimmer, interview.

140. Ibid.

141. C. Mazza and J. F. Kaufeler, "A New Generation of Spacecraft Control System-'SCOS,'" *ESA Bulletin* 56 (November 1988): 23.

142. European Space Agency Annual Report 1994, European Space Agency Headquarters Library, Paris, 95.

143. Wimmer, interview.

144. Debatin, "New Ground Data-Processing System to Support the Agency's Future Satellite Missions," 78–79.

145. Mazza and Kaufeler, "A New Generation of Spacecraft Control System-'SCOS,'" 20.

146. European Space Agency Annual Report 1997, European Space Agency Headquarters Library, Paris, 137.

147. M. Jones, N. C. Head, K. Keyte, and M. Symonds, "SCOS II: ESA's New Generation of Mission-Control System," *ESA Bulletin* 75 (August 1993): 79–84.

148. European Space Agency Annual Report 1999, 45, 68.

149. "SCOS-2000: The Advanced Spacecraft Operations System," European Space Agency Publications.

150. European Space Agency Annual Report 1999, 68.

151. Paolo Ferri, interview by author, Darmstadt, Germany, 20 October 2010.

152. European Space Agency Annual Report 2011, http://www.esa.int/About_Us/ESA_Publications/ESA_Publications_Annual_Report/ESA_Annual_Report_2011, 86.

153. Longdon and David, *ESOC: The European Space Operations Centre.*

154. Warhaut, interview.

155. Howard Nye, interview by author, Paris, France, 12 October 2010.

156. Manfred Warhaut, "ESOC in Context" (PowerPoint presentation, European Space Operations Centre, Darmstadt, Germany, 23 November 2010).

157. Nye, *ESOC*, 2; and Longdon and David, ESOC: *The European Space Operations Centre.*

158. Warhaut, interview; and Wimmer, interview.

159. Schäfer, *How to Survive in Space, Vol. II*, 133.

160. Nye, interview.

161. Nye, *ESOC*, 20.

162. Warhaut, interview.

163. Schäfer, *How to Survive in Space, Vol. II*, 133.

164. Wimmer, interview.

165. Nye, interview.

166. Nye, *ESOC*, 21.

167. Ibid., 3, 29.

168. Warhaut, interview.

169. Longdon and David, *ESOC: The European Space Operations Centre.*

170. "European Space Operations Directorate: . . . constant vigil," 1993, European Space Agency Headquarters Library, Paris.

171. Lacoste, *Europe: Stepping Stones to Space*, 1990, 56–62.

172. European Space Agency Annual Report 1996, European Space Agency Headquarters Library, Paris, 131.

173. P. B. Lemke, "The Geos Ground System," *ESA Bulletin* 9 (May 1977): 53.

174. European Space Agency Annual Report 1994, 95–96.

175. European Space Agency Annual Report 1995, 89, 91.

176. European Space Agency Annual Report 2003, European Space Agency Headquarters Library, Paris, 87.

177. Warhaut, interview.

178. Manfred Warhaut and Andrea Accomazzo, "Venus Express Ground Segment and Mission Operations," *ESA Bulletin* 124 (November 2005): 35–36.

179. Wimmer, interview.

180. Wilkins, "Spacecraft Operations at ESOC," 10.

181. Nye, interview.

182. Hennessy, interview.

183. Warhaut, interview.

184. Nye, interview.

185. Hennessy, interview.

186. Nye, *ESOC*, 17.

187. Gujer and Jabs, "Use of Spacecraft Simulators at ESOC," 41.

188. Stainer and Dworak, "Training in Satellite Ground-System Operations," 67–68.

189. Wimmer, interview.

190. A. Smith et al, "Lost in Space?—ESOC Always Comes to the Rescue," *ESA Bulletin* 117 (February 2004): 56.

191. Ibid., 57.

192. European Space Agency Annual Report 1986, 132.

193. "European Space Operations Directorate: . . . constant vigil"; and Wimmer, interview.

194. Lacoste, 142.

195. Schäfer, *How to Survive in Space, Vol. I*, 168–69.

196. Smith, et al, "Lost in Space?—ESOC Always Comes to the Rescue," 58–59.

197. Nye, *ESOC*, 4.

198. Smith et al, "Lost in Space?—ESOC Always Comes to the Rescue," 59–60.

199. Ibid., 60–63.

200. Battrick, *Supporting Europe's Endeavours in Space*, 1998, 14–16.

201. Schäfer, *How to Survive in Space, Vol. II*, 25.

CHAPTER 4. INTERNATIONAL COOPERATION

1. John Sakss, "NASA and International Space Cooperation," from Thompson and Guerrier, *Space: National Programs and International Cooperation*, 1989, 109.

2. Ibid., 108.

3. Roy Gibson, "Space—New Opportunities for International Ventures," in Hayes, *Space—New Opportunities for International Ventures*, 1980, 3.

4. Bondi, "International Cooperation in Space," 1.

5. Ibid., 6.

6. Zabusky, *Launching Europe*, 1995, 5–6.

7. Gibbons, *The European Space Agency*, 6.

8. Krige and Russo, *A History of the European Space Agency, Volume I*, 338.

9. Wayne C. Thompson, "West Germany's Space Program and the European Effort," in Thompson and Guerrier, *Space: National Programs and International Cooperation*, 1989, 43.

10. ESA Annual Report 1989, 221.

11. Hell, interview.

12. Black and Andrews, *ESOC Services Catalog 2000/2001*.

13. Ibid., 2.

14. Mandfed Warhaut, "ESOC and JPL Cooperation," PowerPoint presentation, European Space Operations Centre, Darmstadt, Germany, 18 October 2010, 11.

15. Longdon, "ESA/ESOC 25 Years."

16. Longdon, "U.S.-European Cooperation in Space Science: A 25-Year Perspective," 1984; and John Rhea, "The Need for More International Cooperation in Space," in Thompson and Guerrier, *Space: National Programs and International Cooperation*, 1989, 113.

17. Ibid., 12.

18. Bondi, "International Cooperation in Space," 2.

19. Longdon, "U.S-European Cooperation in Space Science," 15.

20. Wimmer, interview.

21. Longdon, "U.S-European Cooperation in Space Science," 15.

22. Zabusky, *Launching Europe*, 5.

23. Longdon, "U.S-European Cooperation in Space Science," 15.

24. Kash, *The Politics of Space Cooperation*, 1967, 38.

25. Longdon, "U.S-European Cooperation in Space Science," 15.

26. Mudgway, *Uplink-Downlink*, 89–90.

27. Ibid., 190–91.

28. Oscar E. Anderson, "Financial Arrangements for Visiting Soviet Specialists, Memorandum," 13 March 1974, Box 1325, Apollo Space Program Office Files, ASTP Series, Johnson Space Center History Collection at the University of Houston–Clear Lake.

29. Glynn S. Lunney, interview by Carol Butler, 30 March 1999, transcript, JSC Oral History Collection, 26–29.

30. Pavelka, interview, 25.

31. Frank, interview, 17–18.

32. Lewis, interview, 40–41.

33. Frank, interview, 19.

34. Sakss, "NASA and International Space Cooperation," 108–9.

35. Longdon, "U.S-European Cooperation in Space Science," 11.

36. Committee on International Space Programs, *U.S.-European Collaboration in Space Science*, 1998, 19.

37. Longdon, "U.S.-European Cooperation in Space Science," 14.

38. Bondi, "International Cooperation in Space," 75.

39. Sakss, "NASA and International Space Cooperation," 106.

40. *ESRO/ELDO Bulletin* 17 (February 1972), European Space Agency Headquarters Library, Paris, 36.

41. European Space Agency Annual Report 1992, European Space Agency Headquarters Library, Paris, 191.

42. Ibid.

43. Longdon, "U.S.-European Cooperation in Space Science," 11.

44. Krige and Russo, *A History of the European Space Agency, Volume I*, 75.

45. Committee on International Space Programs, etc., 16.

46. Black and Andrews, *ESOC Services Catalog 2000/2001*, 7.

47. Hell, interview.

48. Fertig, interview.

49. Since ESA has a degree of sovereignty, its employees are granted diplomatic immunity, though ESA reserves the right to disallow said immunity if an employee commits a crime.

50. For more on treaties versus MOUs, please see Gibbons, *The European Space Agency*, 50–75.

51. Gibbons, *The European Space Agency*, 80–83.

52. Longdon, "ESA/ESOC 25 Years."

53. Committee on International Space Programs, etc., 22.

54. Wimmer, interview.

55. Warhaut, "ESOC and JPL Cooperation."

56. Longdon, "International Cooperation in Space," 45–46.

57. Nye, interview.

58. Committee on International Space Programs, etc., 15.

59. Letter from Arnold W. Frutkin to Pierre Auger, 22 July 1965, ESRO 6921, European University Institute Archives, Florence; and MOU between the European Space Research Organization and the National Aeronautics and Space Administration Concerning the Furnishing of Satellite Launching and Associated Services, 11 February 1966, ESRO 6921, European University Institute Archives, Florence.

60. Longdon, "U.S.-European Cooperation in Space Science," 12–13.

61. Committee on International Space Programs, etc., 21; and Longdon, "U.S.-European Cooperation in Space Science," 11.

62. Battrick, *Supporting Europe's Endeavours in Space*, 11–13.

63. Harvey, *The Japanese and Indian Space Programmes*, 2000, 79.

64. Committee on International Space Programs, etc., 23.

65. Wimmer, interview.

66. Letter from Daniel S. Goldin, NASA Administrator to Jean-Martin

Luton, Director-General ESA, 2 November 1992, ESA 15754, European University Institute Archives, Florence.

67. Warhaut, interview; and Ferri, interview.

68. Warhaut, interview.

69. Hell, interview.

70. Warhaut, interview.

71. Harvey, *The Japanese and Indian Space Programmes*, 89.

72. European Space Agency Council and Space Station Working Group, Draft Memorandum of Understanding between NASA and ESA, 11 February 1988, ESA 12238, European University Institute Archives, Florence.

73. For more information on Space Station Freedom, consult Launius, *Space Stations: Base Camps to the Stars*, 2003, 111–41; and Harland and Catchpole, *Creating the International Space Station*, 2002, 88–102 and 113–42.

74. For more information on the International Space Station, consult Launius, *Space Stations*, 175–238; and Harland and Catchpole, *Creating the International Space Station*.

CHAPTER 5. TRACKING NETWORKS

1. Jay A. Holladay, interviewed by Jose Alonso, 9 July, 11 September, and 30 September 1992, transcript, Jet Propulsion Laboratory Archives Oral History Program, 39–40.

2. Corliss, *Histories of the Space Tracking and Data Acquisition Network (STADAN), the Manned Space Flight Network (MSFN), and the NASA Communications Network (NASCOM)*, 1974, 3, 23–24.

3. Ibid., 29–30, 36–37, 42–43.

4. Ibid., 48, 57, 61.

5. Ibid., 86, 95, 100.

6. Ibid., 105–6, 124.

7. Kraft, "Mercury Operational Experience."

8. Corliss, *Histories of the Space Tracking and Data Acquisition Network (STADAN), the Manned Space Flight Network (MSFN), and the NASA Communications Network (NASCOM)*, 145, 147.

9. Kranz, *Failure Is Not an Option*, 142.

10. Corliss, *Histories of the Space Tracking and Data Acquisition Network (STADAN), the Manned Space Flight Network (MSFN), and the NASA Communications Network (NASCOM)*, 180.

11. Ibid., 184–85, 204.

12. Frank, "Flight Control of the Apollo Lunar-Landing Mission," 2.

13. Robert O. Allar and Lorne M. Robinson, "Tracking and Data Relay Satellite System: Space Data System of the 80's," in Hayes, *Space—New Opportunities for International Ventures*, , 1980, 33–35.

14. Lipartito and Butler, 240.

15. "Mission Control Center," NASA Facts, 1986.

16. "TDRS H, I, J," Boeing, http://www.boeing.com/defense-space/space/bss/factsheets/601/tdrs_hij/tdrs_hij.html, accessed 9 February 2012; "Tracking and Data Relay Satellite System (TDRSS)," National Aeronautics and Space Administration, https://www.spacecomm.nasa.gov/spacecomm/programs/tdrss/default.cfm, accessed 9 February 2012; and "TDRS," Encyclopedia Astronautica, http://www.astronautix.com/craft/tdrs.htm, accessed 9 February 2012.

17. "Reference Guide to the International Space Station," November 2010, National Aeronautics and Space Administration, http://www.nasa.gov/pdf/508318main_ISS_ref_guide_nov2010.pdf, accessed 9 February 2012, 94.

18. Allar and Robinson, "Tracking and Data Relay Satellite System: Space Data System of the 80's," 37–40.

19. Amber Hinkle and Dewayne Washington, "All Systems Go for Next Communication Spacecraft," 21 November 2011, National Aeronautics and Space Administration, http://www.nasa.gov/topics/technology/features/tdrs-go.html, accessed 9 February 2012.

20. "Tracking and Data Relay Satellite," Goddard Space Flight Center, http://tdrs.gsfc.nasa.gov/, accessed 17 February 2014.

21. William Pickering, Press Conference at JPL, 7 October 1957, History Collection 3–39, Jet Propulsion Laboratory Library, Pasadena, California.

22. Corliss, *A History of the Deep Space Network*, 1; and C. D. Edwards Jr., C. T. Stelzried, L. J. Deutsch, and L. Swanson, "NASA's Deep-Space Telecommunications Road Map," *TMO Progress Report* 42–136 (15 February 1999): 2.

23. Ibid., 8–12.

24. Ibid., 14–17.

25. Mudgway, *Uplink-Downlink*, 3, 12.

26. Ibid., 14

27. Corliss, *A History of the Deep Space Network*, 34.

28. Holladay, interview, 3.

29. Corliss, *A History of the Deep Space Network*, 43, and Mudgway, *Uplink-Downlink*, 61.

30. Mudgway, *Uplink-Downlink*, 64.

31. Corliss, *A History of the Deep Space Network*, 77–79.

32. Ibid., 99.

33. N. A. Renzetti, "Tracking and Data System Support for *Surveyor* Mission V," Technical Memorandum 33-301, Vol. III, 1 December 1969, Jet Propulsion Laboratory Library, Pasadena, California, 4.

34. Corliss, *A History of the Deep Space Network*, 82–84 and 129, and Mudgway, *Uplink-Downlink*, 45.

35. Robert H. Evans, interview by Jose Alonso, 21 and 23 September 1992, transcript, Jet Propulsion Laboratory Archives Oral History Program, 16.

36. Corliss, *A History of the Deep Space Network*, 110–11.

37. Ibid., 142, 150.

38. Ibid., 175.

39. Mudgway, *Uplink-Downlink*, 45.

40. Corliss, *A History of the Deep Space Network*, 197–200; and Mudgway, *Uplink-Downlink*, 77.

41. Mudgway, *Uplink-Downlink*, 109.

42. Renzetti, "Tracking and Data System Support for *Surveyor* Missions I and II," 2.

43. Renzetti et al, "Tracking and Data System Support for the Mariners Mars 1969 Mission, Planning Phase Through Midcourse Maneuver," 1–2.

44. Mudgway, *Uplink-Downlink*, 32–33.

45. Ibid., 97.

46. Ibid., 207.

47. Ibid., 155–58, 224–26.

48. Ibid., 2–4.

49. Ibid., 198.

50. Ibid., 226–30.

51. Ibid., 221–22.

52. Ibid., 242–45.

53. J. W. Layland and L. L. Rauch, "The Evolution of Technology in the Deep Space Network: A History of the Advanced Systems Program," *TDA Progress Report* 42–130 (15 August 1997): 5.

54. Mudgway, *Uplink-Downlink*, xxxvii.

55. Schäfer, *How to Survive in Space, Vol. II*, 85.

56. First General Report of the European Space Research Organization (1964–1965), III.4, 1–4.

57. Krige and Russo, *A History of the European Space Agency, Volume I*, 59.

58. First General Report of the European Space Research Organization (1964–1965), III.4, 8–9.

59. Longdon and Guyenne, *Twenty Years of European Cooperation in Space*, 95.

60. European Space Agency Annual Report 1977, 147–48.

61. Schäfer, *How to Survive in Space, Vol. II*, 92.

62. Ibid., 90.

63. Longdon and Guyenne, *Twenty Years of European Cooperation in Space*, 95.

64. Schäfer, *How to Survive in Space, Vol. II*, 90.

65. Ibid., 89.

66. Longdon and David, *ESOC: The European Space Operations Centre*.

67. Manfred Bertelsmeier and Gioacchino Buscami, "New Communications Solutions for ESA Ground Stations," *ESA Bulletin* 125 (February 2006): 44–49.

68. Landeau-Constantin, von Weyhe, and Cebers de Sousa, *ESOC*, 14.

69. European Space Agency Annual Report 2008, European Space Agency Headquarters Library, Paris, 94.

70. Schäfer, *How to Survive in Space, Vol. I*, 151–52.

71. European Space Agency Annual Report 2002, European Space Agency Headquarters Library, Paris, 101.

72. Nye, *ESOC*, 25.

73. Wimmer, interview.

74. G. Servoz, "Powering the ESA Network," *ESA Bulletin* 66 (May 1991): 86.

75. Glenn, *John Glenn: A Memoir*, 1999, 264; L. Gordon Cooper Jr., interview by Roy Neal, 21 May 1998, transcript, JSC Oral History Collection, 10; "City of Light—50 Years in Space," 2012, Western Australian Museum, http://museum. wa.gov.au/city-lights, accessed 24 February 2014.

76. Corliss, *Histories of the Space Tracking and Data Acquisition Network (STADAN), the Manned Space Flight Network (MSFN), and the NASA Communications Network (NASCOM)*, 123.

CONCLUSION

1. Fertig, interview.

2. Wimmer estimated the usual ratio of 1:5 ESOC to NASA employees per mission.

3. Christopher C. Kraft Jr., John D. Hodge, and Eugene F. Kranz, "Mission Control for Manned Space Flight," NASA Fact Sheet 170, 23 April 1963, Box 53, General History, General Reference, Johnson Space Center History Collection at the University of Houston–Clear Lake.

4. ESA Annual Report 1991, 151.

5. Kraft, *Flight*, 352.

BIBLIOGRAPHY

ARCHIVES

European Space Agency Headquarters Library, Paris, France.

European Space Operations Centre Library, Darmstadt, Germany.

European University Institute Archives, Florence, Italy.

Jet Propulsion Laboratory Library, Pasadena, California.

Johnson Space Center Collection at University of Houston–Clear Lake, Houston, Texas.

National Aeronautics and Space Administration Headquarters, Washington, DC.

National Archives, Pacific Region, Riverside, California.

National Archives, Southwest Region, Fort Worth, Texas.

Smithsonian National Air and Space Museum Archives, Suitland, Maryland.

PRIMARY SOURCES

Battrick, B., ed. *Supporting Europe's Endeavours in Space: The ESA Directorate of Technical and Operational Support.* Noordwijk, The Netherlands: ESA Publications Division, 1998.

Black, W., and D. Andrews. *ESOC Services Catalog 2000/2001.* Darmstadt, Germany: ESOC External Customer Services Unit, 2000.

"Convention for the Establishment of a European Space Agency." European Space Operations Centre Library, Darmstadt, Germany.

"European Space Operations Directorate: . . . constant vigil." BR 88, European Space Agency Headquarters Library, Paris.

Warhaut, Manfred. "ESA and ESOC Overview." PowerPoint Presentation. European Space Operations Centre, Darmstadt, Germany, 16 August 2010.

———. "ESOC and JPL Cooperation." PowerPoint Presentation. European Space Operations Centre, Darmstadt, Germany, 18 October 2010.

———. "ESOC in Context." PowerPoint Presentation. European Space Operations Centre, Darmstadt, Germany, 23 November 2010

Interviews

Aaron, John. Interview by Kevin M. Rusnak. 26 January 2000. Transcript. JSC Oral History Collection.

Accola, Anne L. Interview by Rebecca Wright. 16 March 2005. Transcript. JSC Oral History Collection.

Aldrich, Arnold D. Interview by Kevin M. Rusnak. 24 June 2000. Transcript. JSC Oral History Collection.

Cooper, L. Gordon, Jr. Interview by Roy Neal. 21 May 1998. Transcript. JSC Oral History Collection.

Dumis, Charles L. Interview by Kevin M. Rusnak. 1 March 2002. Transcript. JSC Oral History Collection.

Evans, Robert H. Interview by Jose Alonso. 21 and 23 September 1992. Transcript. Jet Propulsion Laboratory Archives Oral History Program.

Ferri, Paolo. Interview by author. Darmstadt, Germany. 20 October 2010.

Fertig, Jurgen. Interview by author. Darmstadt, Germany. 20 October 2010.

Frank, M. P. "Pete," III. Interview by Doyle McDonald. 19 August 1997. Transcript. JSC Oral History Collection.

Greene, Jay H. Interview by Sandra Johnson. 10 November 2004. Transcript. JSC Oral History Collection.

Greene, Jay H. Interview by Sandra Johnson. 8 December 2004. Transcript. JSC Oral History Collection.

Griffin, Gerald D. Interview by Doug Ward. 12 March 1999. Transcript. JSC Oral History Collection.

Harlan, Charles S. Interview by Kevin M. Rusnak. 14 November 2001. Transcript. JSC Oral History Collection.

Hasbrook, Annette P. Interview by Jennifer Ross-Nazzal. 21 July 2009. Transcript. JSC Oral History Collection.

Hell, Wolfgang. Interview by author. Darmstadt, Germany. 20 October 2010.

Hennessy, Leo. Interview by author. Paris, France. 12 October 2010.

Hodge, John D. Interview by Rebecca Wright. 18 April 1999. Transcript. JSC Oral History Collection.

Holladay, Jay A. Interviewed by Jose Alonso. 9 July, 11 September, and 30 September 1992. Transcript. Jet Propulsion Laboratory Archives Oral History Program.

Kranz, Eugene F. Interview by Roy Neal. 19 March 1998. Transcript. JSC Oral History Collection.

Kranz, Eugene F. Interview by Rebecca Wright. 8 January 1999. Transcript. JSC Oral History Collection.

Kranz, Eugene F. Interview by Roy Neal. 28 April 1999. Transcript. JSC Oral History Collection.

Kranz, Eugene F. Interview by author. 22 November 2006. Transcript. Skylab Oral History Project, University of North Texas Oral History Program.

Kranz, Gene. Interview by Jo Jeffrey Kluger. 29 May 1992. Transcript. Kranz, Eugene F. (NASA-Bio.) 1243. National Aeronautics and Space Administration Headquarters Archives, Washington, DC.

Lewis, Charles. Interview by author. 22 September 2006. Transcript. Skylab Oral History Project, University of North Texas Oral History Program.

Loe, T. Rodney. Interview by Carol L. Butler. 30 November 2001. Transcript. JSC Oral History Collection.

Loe, T. Rodney. Interview by Carol L. Butler. 7 November 2001. Transcript. JSC Oral History Collection.

Lunney, Glynn S. Interview by Carol Butler. 30 March 1999. Transcript. JSC Oral History Collection.

Lunney, Glynn S. Interview by Carol Butler. 28 January 1999. Transcript. JSC Oral History Collection.

Lunney, Glynn S. Interview by Roy Neal. 9 March 1998. Transcript. JSC Oral History Collection.

McClure, Jim, and Ron Sharp. Interview by author. Pasadena, California. 3 September 2010.

Nye, Howard. Interview by author. Paris, France. 12 October 2010.

Pavelka, Edward L., Jr. Interview by Carol Butler. 26 April 2001. Transcript. JSC Oral History Collection.

Pavelka, Edward L., Jr. Interview by Carol Butler. 9 March 2001. Transcript. JSC Oral History Collection.

Purser, Paul. Interview by Robert Merrifield. 17 May 1967. Transcript. Manned Spacecraft Center (MSC) History Interviews Kr-Z, Folder 15, Box 2, 18994. National Aeronautics and Space Administration Headquarters Archives, Washington, DC.

Warhaut, Manfred. Interview by author. Darmstadt, Germany. 19 October 2010.

Webb, James. Interview. 15 March 1985. Transcript. Glennan-Webb-Seamans Project Interviews. National Air and Space Museum Archives, Suitland, MD.

Weitz, Paul J. Interview by Rebecca Wright. 26 March 2000. Transcript. JSC Oral History Collection.

Wimmer, Wolfgang. Interview by author. Darmstadt, Germany. 21 October 2010.

SECONDARY SOURCES

Books

Bergreen, Laurence. *Voyage to Mars: NASA's Search for Life Beyond Earth*. New York: Riverhead Books, 2000.

Black, W., and D. Andrews. *ESOC Services Catalog 2000/2001*. Darmstadt, Germany: ESOC External Customer Services Unit, 2000.

Bondi, Hermann. "International Cooperation in Space." In *International Cooperation in Space Operations and Exploration*, edited by Michael Cutler, 1–6. Tarzana, CA: American Astronautical Society, 1971.

Bonnet, Roger M., and Vittorio Manno. *International Cooperation in Space: The Example of the European Space Agency*. Cambridge, MA: Harvard University Press, 1994.

Chaikin, Andrew. *A Man on the Moon, Vol. I: One Giant Leap*. Alexandria, VA: Time-Life Books, 1994.

———. *A Man on the Moon, Vol. II: The Odyssey Continues*. Alexandria, VA: Time-Life Books, 1994.

Committee on International Space Programs, National Research Council, and European Space Science Committee, European Space Foundation. *U.S.-European Collaboration in Space Science*. Washington, DC: National Academy Press, 1998.

Conrad, Nancy, and Howard A. Klausner. *Rocket Man: Astronaut Pete Conrad's Incredible Ride to the Moon and Beyond*. New York: New American Library, 2005.

Corliss, William R. *Histories of the Space Tracking and Data Acquisition Network (STADAN), the Manned Space Flight Network (MSFN), and the NASA Communications Network (NASCOM)*. National Aeronautics and Space Administration, 1974.

———. *A History of the Deep Space Network*. Washington, DC: National Aeronautics and Space Administration, 1976.

Dethloff, Henry C. . . . *Suddenly, Tomorrow Came: A History of the Johnson Space Center*. Houston: NASA Johnson Space Center, 1993.

Dethloff, Henry C., and Ronald A. Schorn. *Voyager's Grand Tour: To the Outer Planets and Beyond*. Old Saybrook, CT: Konecky & Konecky, 2003.

Durant, Frederick C., III, and George S. James, eds. *First Steps Towards Space: Proceedings of the First and Second History Symposia of the International Academy of Astronautics at Belgrade, Yugoslavia, 26 September 1967, and New York, U.S.A., 16 October 1968*. Washington: Smithsonian University Press, 1974.

Glenn, John, with Nick Taylor. *John Glenn: A Memoir*. New York: Bantam, 1999.

Hall, R. Cargill, ed. *Essays on the History of Rocketry and Astronautics: Proceedings of the Third Through the Sixth History Symposia of the International Acad-*

emy of Astronautics, Vol. II. Washington: National Aeronautics and Space Administration, Scientific and Technical Information Office, 1977.

Hansen, James R. *First Man: The Life of Neil A. Armstrong.* New York: Simon and Schuster, 2005.

Harland, David M., and John E. Catchpole. *Creating the International Space Station.* London: Springer-Praxis, 2002.

Harvey, Brian. *The Japanese and Indian Space Programmes: Two Roads into Space.* London: Springer-Praxis, 2000.

Hayes, William C., Jr., ed. *Space—New Opportunities for International Ventures.* 17th Goddard Memorial Symposium, Vol. 49. San Diego: American Aeronautical Society, 1980.

Jansson, André, and Amanda Lagerkvist, eds. *Strange Spaces: Explorations into Mediated Obscurity.* Surrey, England: Ashgate Publishing Limited, 2009.

Kash, Don E. *The Politics of Space Cooperation.* West Lafayette, IN: Purdue University Studies, 1967.

Koppes, Clayton R. *JPL and the American Space Program: A History of the Jet Propulsion Laboratory.* New Haven: Yale University Press, 1982.

Kraft, Chris. *Flight: My Life in Mission Control.* New York: Dutton, 2001.

Kranz, Gene. *Failure Is Not an Option: Mission Control from Mercury to Apollo 13 and Beyond.* New York: Simon and Schuster, 2000.

Krige, John, and Arturo Russo. *A History of the European Space Agency, 1958–1987: Volume I: The Story of ESRO and ELDO, 1958–1973.* Noordwijk, The Netherlands: ESA Publications Division, 2000.

Lacoste, Beatrice. *Europe: Stepping Stones to Space.* Bedfordshire, UK: Orbic, 1990.

Landeau-Constantin, Jocelyne, Bernhard von Weyhe, and Nicola Cebers de Sousa, compilers. *ESOC.* Noordwijk, The Netherlands: ESA Publications Division, 2007.

Launius, Roger D. *Space Stations: Base Camps to the Stars.* Old Saybrook, CT: Konecky & Konecky, 2003.

Levin, Miriam R., ed. *Cultures of Control.* Amsterdam: Harwood Academic Publishers, 2000.

Lipartito, Kenneth, and Orville R. Butler. *A History of the Kennedy Space Center.* Gainesville, FL: University Press of Florida, 2007.

Longdon, Norman, and V. David, compiled and ed. *ESOC: The European Space Operations Centre.* Noordwijk, The Netherlands: ESA Publications Division, 1988.

Longdon, Norman, and Duc Guyenne, eds. *Twenty Years of European Cooperation in Space: An ESA Report.* Paris: European Space Agency Scientific and Technical Publications Branch, 1984.

McCurdy, Howard E. *Faster, Better, Cheaper: Low-Cost Innovation in the U.S. Space Program.* Baltimore: The Johns Hopkins University Press, 2001.

Mishkin, Andrew. *Sojourner: An Insider's View of the Mars Pathfinder Mission.* New York: Berkley Books, 2004.

Mudgway, Douglas J. *Uplink-Downlink: A History of the Deep Space Network 1957–1997.* Washington: National Aeronautics and Space Administration, 2001.

Murray, Charles, and Catherine Bly Cox. *Apollo: The Race to the Moon.* New York: Simon and Schuster, 1989.

NASA Public Affairs. *The Kennedy Space Center Story.* Cape Canaveral, FL: Kennedy Space Center, 1991.

National Aeronautics and Space Administration. *Kennedy Space Center Story.* Kennedy Space Center, FL: John F. Kennedy Space Center, 1972.

Nye, Howard. *ESOC: European Space Operations Centre.* Noordwijk, The Netherlands: ESA Publications Division, 1996.

Schäfer, Madeleine. *How to Survive in Space! (A Light-Hearted Chronicle of ESOC), Volume I (1963–1986).* Darmstadt: European Space Agency, 1997.

———. *How to Survive in Space! (A Light-Hearted Chronicle of ESOC), Volume II (1987–1997).* Darmstadt: European Space Agency, 1997.

Shayler, David J. *Skylab: America's Space Station.* London: Springer, 2001.

Thompson, Wayne C., and Steven W. Guerrier, eds. *Space: National Programs and International Cooperation.* Boulder, CO: Westview Press, 1989.

von Kármán, Theodore, with Lee Edson. *The Wind and Beyond: Theodore von Kármán, Pioneer in Aviation and Pathfinder in Space.* Boston: Little, Brown and Company, 1967.

Westwick, Peter J. *JPL and the American Space Program, 1976–2004.* New Haven: Yale University Press, 2007.

Zabusky, Stacia E. *Launching Europe: An Ethnography of European Cooperation in Space Science.* Princeton: Princeton University Press, 1995.

Journal Articles

Canby, Thomas Y. "Skylab, Outpost on the Frontier of Space." *National Geographic* 146, no. 4 (October 1974): 441–69.

Edwards, C. D., Jr., C. T. Stelzried, L. J. Deutsch, and L. Swanson. "NASA's Deep-Space Telecommunications Road Map." *TMO Progress Report* 42–136 (15 February 1999): 1–20.

Kearney, Michael W., III. "The Evolution of the Mission Control Center." *Proceedings of the IEEE* 75, no. 3 (March 1987): 399–416.

Layland, J. W., and L. L. Rauch. "The Evolution of Technology in the Deep Space Network: A History of the Advanced Systems Program." *TDA Progress Report* 42–130 (15 August 1997): 1–44.

Long, James E. "To the Outer Planets." *Astronautics and Aeronautics* 7 (June 1969): 32–47.

Longdon, John M. "U.S.-European Cooperation in Space Science: A 25-Year Perspective." *Science* 223, no. 4631 (6 January 1984): 11–16.

Viorst, Milton. "The Bitter Tea of Dr. Tsien." *Esquire* 68 (September 1967): 125–29, 168.

Other Publications

Clark, Evert. "New NASA Center Making Its Debut." *New York Times*, 3 June 1965, 21.

Dunn, Louis, W. B. Powell, and Howard Seifert. "Heat Transfer Studies Relating to Rocket Power Plant Development." Proceedings of the Third Anglo-American Conference, 1951, 271–328.

Gibbons, Margaret Ann. *The European Space Agency: Cooperation and Competition in Space.* Geneva: Graduate Institute of International Studies, 1986.

"SCOS-2000: The Advanced Spacecraft Operations System." European Space Agency Publications.

Websites

"City of Light—50 Years in Space." 2012. Western Australian Museum. http://museum.wa.gov.au/city-lights (accessed 24 February 2014).

Hinkle, Amber, and Dewayne Washington. "All Systems Go for Next Communication Spacecraft." 21 November 2011. National Aeronautics and Space Administration. http://www.nasa.gov/topics/technology/features/tdrs-go.html (accessed 9 February 2012).

NASA Office of the General Counsel. "The National Aeronautics and Space Act, Sec. 20102 a." NASA. http://www.nasa.gov/offices/ogc/about/space_act1.html (accessed 16 January 2012).

"Neutral Buoyancy Laboratory." National Aeronautics and Space Administration. http://dx12.jsc.nasa.gov/site/index.shtml (accessed 17 January 2012).

"Reference Guide to the International Space Station." November 2010. National Aeronautics and Space Administration. http://www.nasa.gov/pdf/508318main _ISS_ref_guide_nov2010.pdf (accessed 9 February 2012).

"TDRS." Encyclopedia Astonautica. http://www.astronautix.com/craft/tdrs.htm (accessed 9 February 2012).

"TDRS H, I, J." Boeing. http://www.boeing.com/defense-space/space/bss/factsheets/601/tdrs_hij/tdrs_hij.html (accessed 9 February 2012).

"Tracking and Data Relay Satellite." Goddard Space Flight Center. http://tdrs.gsfc.nasa.gov/. (accessed 17 February 2014).

"Tracking and Data Relay Satellite System (TDRSS)." National Aeronautics and Space Administration. https://www.spacecomm.nasa.gov/spacecomm/programs/tdrss/default.cfm (accessed 9 February 2012).

Webster, Guy. "NASA's Spirit Rover Completes Mission on Mars." 25 March 2011. National Aeronautics and Space Administration. http://www.nasa.gov/mission_pages/mer/news/mer20110525.html (accessed 9 February 2012).

Webster, Guy, and Dwayne Brown. "NASA Concludes Attempts to Contact Mars Rover Spirit." 24 May 2011. Jet Propulsion Laboratory, California Institute of Technology. http://www.jpl.nasa.gov/news/news.cfm?release=2011-156&cid=release_2011-156 (accessed 9 February 2012).

Webster, Guy, and Dwayne Brown. "Now a Stationary Research Platform, NASA's Mars Rover Spirit Starts a New Chapter in Red Planet Scientific Studies." 26 January 2010. Jet Propulsion Laboratory, California Institute of Technology. http://marsrover.nasa.gov/newsroom/pressreleases/20100126a.html (accessed 9 February 2012).

INDEX

Page numbers in *italics* refer to illustrations.

MICHAEL PETER JOHNSON completed his PhD in the history of technology at Auburn University in 2012. He taught at Grand Valley State University before joining the seminary for the Roman Catholic Archdiocese of Galveston-Houston.